Easy Mathematics for Biologists

Easy Mathematics for Biologists

Peter C. Foster
Department of Applied Biology
University of Central Lancashire
Preston
UK

harwood academic publishers
Australia • Canada • China • France • Germany • India • Japan
Luxembourg • Malaysia • The Netherlands • Russia • Singapore
Switzerland

Printed in India.

Amsteldijk 166
1st Floor
1079 LH Amsterdam
The Netherlands

British Library Cataloguing in Publication Data

A catalogue record for this book is available from the British Library.

ISBN 90-5702-338-5 (hardcover)

Contents

Preface

Over the past decade or so I have become increasingly concerned that many students on B.Sc. and HND courses in the biological sciences have great difficulty with numerical calculations relating to aspects of biology, whether it is with both the mathematics and the applications, or just a problem applying the mathematics to a particular situation. While many mathematics textbooks exist, these tend to be either too general and lacking in relevant examples, or too advanced, covering topics which are not required by most biology students. (I am using the term 'biology' to cover all the biologically-related disciplines such as ecology, zoology, botany, biochemistry, physiology, and microbiology.)

This book is intended to be used primarily by such first year students to develop their skills in this area. Rather than provide a conventional textbook, I have chosen to write this as a self-contained workbook. The aim is for most students to be able to work through the book independently of any mathematics lectures. Alternatively, it could be used as part of a variety of courses, whether skills-based or subject-based. In my own university I have used it in place of lectures as the basis of a numeracy skills module for a variety of biology students. The students also received some tutorial help.

The book is arranged in chapters which lead from the basic mathematical ideas of fractions, decimals and percentages, through ratio and proportion, multipliers, exponents and logarithms, to straight line graphs, and finally to graphs that are not straight lines, and their transformation. The associated applications covered are the types of problem most commonly encountered in degree and HND courses in the biological sciences and include concentrations and dilutions, changing units, pH, and linear and non-linear rates of change. Each chapter starts with an explanation of the mathematical concepts covered together with worked examples. The main concepts or definitions to be remembered are highlighted. Worked examples of the applications in biology follow the 'pure' mathematics sections. The applied examples obviously require some understanding of the science as well as the mathematics; I have tried to explain the science to the extent necessary to tackle the problems. At the end of each chapter are numerous examples of both pure and applied problems. Answers to the problems are at the end of the book.

I am indebted to a number of people who have made helpful suggestions and encouraged the development of the book. Peter Robinson, Philip Roberts and Sally Foster read drafts of the text and made useful suggestions. Many of the students who used a draft version also either provided helpful comments or showed me what they found most difficult. Sally Foster also pointed out ambiguities, suggested alternative ways of presenting some arguments, and identified typographical errors and errors of grammar and punctuation. Peter Robinson helped with any information technology problems I had.

1 INTRODUCTION

1.1 HOW TO USE THIS BOOK

You are probably using this book because you have found you have some difficulty with numerical calculations relating to aspects of biology. You are not alone! Many students of the biological sciences have the same problem. Nevertheless it *is* important that you develop the ability to carry out various mathematical procedures and to understand quantitative information being produced as evidence to support some argument or hypothesis. The contents of this book represent the type of problems most commonly encountered in degree and HND courses in the biological sciences. Mastery of these will mean you should improve your confidence in understanding many parts of your course.

The book is arranged in chapters which are sequential in that I believe you need to master the contents of one before you will be able to fully understand the next. Each chapter or part of a chapter starts with an explanation of the mathematical concepts covered, follows with examples that are purely mathematical, and ends with examples that are applications of the mathematics in biology. You may feel competent at doing the calculations in the first chapter(s) and be tempted to ignore them. You may have difficulties with both the mathematics and the applications, or just have a problem applying the mathematics to a particular situation. Whatever the case, I suggest you read each chapter and try out the 'pure' mathematics examples marked'*'. If you get these correct, try all of the applied problems. If you also get these correct, move on to the next chapter. If you do not get the right answers, reread the explanations and try the rest of the 'pure' problems, and then repeat your attempts at the applied problems.

Answers to the problems are at the end of the book. Try to resist the temptation to look at the answers before committing yourself to an answer in writing. It is very easy to fool yourself that you can do a problem if you look at the answer first!

1.2 UNDERSTANDING A PROBLEM – WHAT YOU NEED TO KNOW

It is quite likely that you are able to do many of the 'pure' mathematical problems in the following chapters, but that your problem is working out how to manipulate the data you are presented with in a real situation. The following example of a typical situation in biochemistry may help to illustrate how to understand the problem and dissect out the relevant parts.

Suppose you are conducting an experiment to measure the rate of an enzyme-catalysed reaction. (An enzyme speeds up a specific chemical reaction.) You want to find out how the rate of reaction changes if you vary the concentration of one of the reactants, called PEP. You know you will want to measure the rate at about 9 different concentrations, and that the concentrations have to start at about 0.2 mmol/l and go down by a factor of about 100. The final volume of the solution in which you are measuring the rate is 2.0 ml. What volumes and concentrations of PEP are you going to use?

What you need to understand is:

- what is meant by concentration, by volume, and by amount, and their interrelationships;
- what 0.2 mmol/l means;
- how to convert multiples of one unit to another;
- that there is a physical limit to the volume of a solution you can pipette, determined by your equipment.

What you do *not* need to know to do this problem is anything about the enzyme or reaction, so don't worry about these. The first three above require that you understand the mathematical ideas of fractions, decimals, proportions, exponents and units. These are covered in the first few chapters and, together with straight line graphs, are the most important concepts you need to understand and be able to use. The later chapters cover other topics which are also important, but less basic.

2 MEASUREMENTS AND UNITS: FRACTIONS, DECIMALS, AND PERCENTAGES

2.1 MEASUREMENTS AND UNITS

The quantities you will come across most often are length, mass, amount of substance, time and temperature. Measurements of these consist of a number and a unit. The number expresses the ratio of the measured quantity to a fixed standard, and the unit is the name of the standard. For example, if a piece of wood is 3 metres long it is 3 times the length of the standard used to define 1 metre. **It is vitally important that you use units correctly and especially that you write down the correct unit after calculating a numerical answer**. (Some measurements do not have units. This may be because they are ratios of two measurements with the same units, or because they are a particular kind of mathematical transformation of a measurement called a logarithm. Logarithms are explained in Chapter 6.)

A large number of different units could be used. Distance could be measured in metres, inches, hands, feet, furlongs, miles, or light-years! The accepted convention for use in science is the 'SI' system in which there are 'base' units and 'derived' units each of which has a specified abbreviation or symbol. The base units for length, mass, amount of substance, time and temperature are metre (m), kilogram (kg), mole (mol), second (s) and kelvin (K) respectively. Derived units are made up from combinations of base units; for example, the unit of energy, the joule, is metres squared times kilograms divided by seconds squared.

It is very important that you write down the appropriate unit with the number and that you use the correct abbreviation or symbol. Otherwise, you may confuse the reader or the information you are giving may be meaningless.

Measurements should be written with the number separated from the unit by a space, and the unit should be singular and without the full stops normally used to show an abbreviation. For example, a mass of 6 kilograms is written as 6 kg and not as 6kg or 6 kg. or 6kgs. Where units are combined there is a space between them. For example, a speed of 6 metres per second is written as $6\ \mathrm{m\ s^{-1}}$, and not as $6\ \mathrm{ms^{-1}}$ or $6\ \mathrm{m.s^{-1}}$. ($\mathrm{s^{-1}}$ is shorthand for 'per second': $6\ \mathrm{m\ s^{-1}}$ is the same as writing 6 m/s.) An area of 4 square metres is written as $4\ \mathrm{m^2}$ and not as $4\mathrm{m^2}$ and a volume of 2 cubic metres is written as $2\ \mathrm{m^3}$. (Note: $\mathrm{m^2}$ is shorthand for m $\times$ m, square metres, the area of a square of side 1 metre. Similarly, m^3 is shorthand for m × m × m, metres cubed, the volume of a cube of side 1 metre.)

Although using SI makes calculations easier in some ways, it is not perfect, and most biologists are sensible enough to use other units if there is a good reason. For example, when reporting changes over days or years it would be silly to use seconds as the unit. Particularly common is the use of the litre as a measure of volume for the reasons that it is a convenient size; laboratory glassware is calibrated using it; in most countries (and increasingly in the UK) people are familiar with it; and it is readily convertible to the SI units. The litre is normally abbreviated as 'l', but many American texts use 'L'. There are one thousand litres in a cubic metre. (A litre is the volume of a cube whose sides are one tenth of a metre by one tenth of a metre. One tenth of a metre is also known as a decimetre (dm), so a litre is also the same as a cubic decimetre.)

We will come back to more about units in later sections, but now I want to move on to ways of expressing quantities that are not whole numbers.

2.2 FRACTIONS

A fraction is a whole number divided by another whole number other than 0, e.g. $^3/_5$. (Note that the symbol '/' means 'divided by' whether used in a simple fraction or a more complex situation.)

The number above the line (3) is called the **numerator** and that below the line (5) is called the **denominator**.

Two fractions are equal if one can be converted to the other by multiplying or dividing both the numerator and the denominator by the same number. These are called **equivalent fractions**.

Example 2.1 $^5/_8 = {}^{10}/_{16}$
because $5 \times 2 = 10$
and $8 \times 2 = 16$

Example 2.2 Are $^9/_{21}$ and $^3/_7$ the same?

Yes, because 9 divided by 3 = 3
and 21 divided by 3 = 7.

To **add fractions** you need to get the **denominators** the same.

Example 2.3 $^5/_8 + {}^7/_{16} = ?$

Converting $^5/_8$ to its equivalent fraction $^{10}/_{16}$

gives both fractions the same denominator 16
so $^{10}/_{16} + ^{7}/_{16} = ^{17}/_{16}$ or $1^{1}/_{16}$

Example 2.4 $^{3}/_{11} + ^{5}/_{8} = ?$

Converting both fractions to have the denominator 88
$^{3}/_{11} = (3 \times 8)/(11 \times 8) = ^{24}/_{88}$
$^{5}/_{8} = (5 \times 11)/(8 \times 11) = ^{55}/_{88}$
therefore $^{3}/_{11} + ^{5}/_{8} = ^{24}/_{88} + ^{55}/_{88} = ^{79}/_{88}$

Example 2.5 $2\,^{3}/_{4} + ^{4}/_{5} = ?$

Converting both to have 20 as the denominator
$2\,^{3}/_{4} = 2 + (3 \times 5)/(4 \times 5) = 2\,^{15}/_{20}$
$^{4}/_{5} = (4 \times 4)/(5 \times 4) = ^{16}/_{20}$
so $2\,^{3}/_{4} + ^{4}/_{5} = 2\,^{15}/_{20} + ^{16}/_{20}$
$= 2\,^{31}/_{20} = 3\,^{11}/_{20}$
or, $2\,^{3}/_{4} = ^{11}/_{4} = (11 \times 5)/(4 \times 5) = ^{55}/_{20}$
$^{4}/_{5} = ^{16}/_{20}$
$^{55}/_{20} + ^{16}/_{20} = ^{71}/_{20} = 3\,^{11}/_{20}$

Similarly, to **subtract fractions**, the denominators must be the same.

Example 2.6 $5 - ^{4}/_{9} = ?$

Converting to denominator 9
$^{(5\times9)}/_{9} - ^{4}/_{9} = ^{45}/_{9} - ^{4}/_{9}$
$= ^{41}/_{9} = 4\,^{5}/_{9}$
or, $5 - ^{4}/_{9} = 4 + (1 - ^{4}/_{9}) = 4\,^{5}/_{9}$

When **multiplying** fractions, multiply the numerators; then multiply the denominators; then if possible reduce by dividing both the numerator and denominator by any whole number that will leave a whole number. You may also be able to cancel numbers appearing as factors of both the numerator and denominator.

Example 2.7 $^{7}/_{9} \times ^{5}/_{8} = \dfrac{7 \times 5}{9 \times 8} = ^{35}/_{72}$

Example 2.8 $^{7}/_{9} \times ^{3}/_{8} = \dfrac{7 \times 3}{9 \times 8} = ^{21}/_{72} = ^{7}/_{24}$

Note that in this example you could have reduced earlier:

$$^{7}/_{9} \times ^{3}/_{8} = \frac{7 \times 3}{9 \times 8} = \frac{7 \times 1}{3 \times 8} = ^{7}/_{24}$$

Example 2.9 $\frac{4}{7} \times 2\frac{3}{4} = ?$

As $2\frac{3}{4} = \frac{11}{4}$

$\frac{4}{7} \times 2\frac{3}{4} = \frac{4}{7} \times \frac{11}{4}$

Cancelling the 4's gives $\frac{11}{7}$

therefore $\frac{4}{7} \times 2\frac{3}{4} = \frac{11}{7}$

Cancelling at the beginning is particularly useful when you are multiplying or dividing several fractions or a complex fraction:

Example 2.10 $\frac{1}{10} \times \frac{3}{7} \times \frac{100}{18} \times \frac{21}{10} = ?$

This is the same as

$$\frac{1}{10} \times \frac{3}{7} \times \frac{10 \times 10}{3 \times 3 \times 2} \times \frac{7 \times 3}{10} = \frac{1}{2}$$

The way that you cancel does not matter so long as whatever you do to the numerator, you do to the denominator.

To **divide fractions**, multiply the first fraction by the reciprocal (i.e. by the fraction turned upside down) of the second and then reduce or cancel wherever possible.

Example 2.11 The reciprocal of $\frac{3}{4} = \frac{4}{3}$

Example 2.12 $\frac{7}{9} \div \frac{3}{4} = \frac{7}{9} \times \frac{4}{3} = \frac{28}{27} = 1\frac{1}{27}$

Example 2.13 $\frac{7}{9} \div \frac{14}{3} = \frac{7}{9} \times \frac{3}{14} = \frac{1 \times 1}{3 \times 2} = \frac{1}{6}$

Example 2.14 $\frac{2}{3} \div 2\frac{1}{4} = \frac{2}{3} \div \frac{9}{4} = \frac{2}{3} \times \frac{4}{9} = \frac{8}{27}$

2.3 DECIMALS

Although many calculations involve fractions, the answer is usually presented in a decimal form rather than as a fraction, e.g. as 0.375 g, not $\frac{3}{8}$ g.

Remember that in decimals, each position going further to the right of the decimal point is one tenth of the previous one, and each to the left is 10 times bigger.

(hundreds) (tens) (units) . (tenths) (hundredths) (thousandths)

so 475.356 could be written as the fractions 475 $^{356}/_{1000}$ or as $\frac{475356}{1000}$

When **multiplying decimals** you can first ignore the decimal points and multiply the numbers. Then count the number of digits to the right of the decimal points in the original numbers. This will be the number of digits to the right of the decimal point in the answer.

Example 2.15 32.6 × 7.53 = ?

```
  32.6    1 digit to right of decimal point
× 7.53    2 digits to right of decimal point
   978
  1630
  2282
245.478   3 digits to right of decimal point
```

To **divide a decimal** by a decimal, keep multiplying both numbers by 10 until the divisor (the number you are dividing by) is a whole number. Then divide as for a whole number. The decimal point is above the decimal point in the number being divided.

Example 2.16 Divide 71.3 by 3.52

```
        20.25 etc.
352 ⌈ 7130.
      704
        90
         0
        900
        704
        1960
        1760

         200 etc.
```

In practice, you will almost certainly prefer to use a calculator for these kinds of calculations. Whether you do or not, you should get into the habit of making an estimate of the answer as a check on your calculation. In the above example, if you were to use a calculator, you should first estimate the answer by rounding the numbers: $^{71.3}/_{3.52}$ is between $^{70}/_{4}$ and $^{70}/_{3}$. $^{70}/_{4}$ is 17.5 and $^{70}/_{3}$ is about 23.3, so the answer should be between these, i.e. about 20.

2.4 PERCENTAGES

Percentages are a particular kind of fraction where the denominator is 100 but is indicated by the % sign.

Example 2.17 $^5/_{100} = 5\%$; $^{47}/_{100} = 47\%$

To **convert any fraction or decimal into a percentage**, multiply by 100.

Example 2.18 $\frac{7}{8} = \frac{7 \times 100}{8} = 87.5\%$

Example 2.19 $0.37 = 0.37 \times 100 = 37\%$

To **convert a percentage to a fraction or a decimal**, divide by 100.

Example 2.20 $32\% = {}^{32}/_{100} = 0.32$

To **calculate a percentage**, first calculate the fraction, then multiply by 100.

Example 2.21 In a group of 42 students, 10 are male. What percentage are male?

$^{10\ \text{male}}/_{42\ \text{total}} \times 100 = 23.8\%$

To **apply a percentage to a number**, multiply by the number % and divide by 100.

Example 2.22 If 51% of the population of the UK is female, and the total population is 56 million, what is the female population?

$^{51}/_{100} \times 56\ 000\ 000 = 28\ 560\ 000 = 28.56$ million

You will often come across a **percentage increase or decrease**.

Example 2.23 The number of unemployed fell from 2 312 000 to 2 112 000. What percentage fall is this?

Reduction in unemployed is $2\ 312\ 000 - 2\ 112\ 000 = 200\ 000$

$\%\ \text{reduction} = {}^{\text{reduction}}/_{\text{original}} \times 100$

$^{200\ 000}/_{2\ 312\ 000} \times 100 = 8.7\%$

Example 2.24 The number of unemployed rose by 1.7% last month from 2 312 000. What is the new figure?

new figure = old number + increase

increase $= 2\ 312\ 000 \times {}^{1.7}/_{100} = 39.304$

total $= 2\ 312\ 000 + 39\ 304 = 2\ 351\ 304$

An alternative and quicker way of doing this is to realise that the original number is 100%, so the new number will be (100% + the percentage increase) $\times$ the original number.

$$2\ 312\ 000 \times {}^{(100+1.7)}/_{100} =$$
$$2\ 312\ 000 \times {}^{101.7}/_{100} = 2\ 351\ 304$$

Example 2.25 If the number falls by 1.7% of this new figure next month, what will the number be then?

new number = old number – decrease

decrease = $2\ 351\ 304 \times {}^{1.7}/_{100} = 39\ 972$
total = $2\ 351\ 304 - 39\ 972 = 2\ 311\ 332$

or $2\ 351\ 304 \times {}^{(100-1.7)}/_{100} =$
$2\ 351\ 304 \times {}^{98.3}/_{100} = 2\ 311\ 332$

Note that this answer is not the same as the original number. A 1.7% increase followed by a 1.7% decrease does not cancel out because the number you are taking the percentage of has changed. It is important in problems that you are clear to what number a percentage refers.

One way of describing the concentration of a chemical solution is to say the weight of chemical in a certain volume of solution. For example, the appropriate concentration for a pesticide might be achieved by dissolving a sachet containing 50 g in a volume of 1 gallon of water. If the final volume is also 1 gallon, the concentration is therefore 50 g per gallon.

In biology, concentrations of solutions are often described in terms of weight per volume, usually when the molecular weight of the solute is not known, e.g. protein. It is usual to quote the concentration as a number of g, (x g), per 100 ml and this is abbreviated to x %(w/v). The (w/v) stands for weight/volume, indicating it is the weight (g) divided by the volume (100 ml). (Note that 1 ml is 1 millilitre which is one thousandth of a litre.)

Example 2.26 How would you prepare 250 ml of a 4%(w/v) solution of NaCl in water?

As 4%(w/v) is 4 g/100 ml
you need 250 ml of a ${}^{4\,g}/_{100\,ml}$ solution
$= {}^{250\times4}/_{100} = {}^{1000\,g}/_{100} = 10\,g$
so dissolve 10 g of NaCl in water,
and make up to 250 ml final volume.

Example 2.27 If 5 g of NaCl is dissolved, then diluted with water to make 250 ml, what is the %(w/v) concentration of NaCl in the solution?

$$\text{Concentration} = {}^{5\,g}/_{250\,ml} = 0.02\,g/ml$$
$$\%\,(w/v) = g/100\,ml$$
$$\text{so } 0.02\,g/ml \times 100 = 2\,g/100\,ml = 2\%(w/v)$$

It is also possible to describe concentrations as x %(v/v) or x %(w/w). For example, a 50% (v/v) ethanol solution is made by mixing 50 ml ethanol with 50 ml water. A 5% (w/w) sodium chloride solution is made by mixing 5 g sodium chloride and 95 g water. Note that this is not quite the same thing as 5% (w/v) which is 5 g sodium chloride in 100 ml solution final volume, so you would need to dissolve the salt and make up to volume in a volumetric flask.

Example 2.28 How would you prepare 15 ml of a 5%(v/v) solution of ethanol in water?

You need $15\,ml \times 5\,ml/100\,ml$
$= 75\,ml/100 = 0.75\,ml$ of ethanol
So add 0.75 ml of ethanol to 14.25 ml of water.

Example 2.29 What is the %(v/v) concentration of a solution of ethanol made by mixing 25 ml ethanol with 225 ml of water?

$\%(v/v) = ml/100\,ml$
Total volume $= 25 + 225 = 250\,ml$
so $\%(v/v) = {}^{25}/_{250} \times 100 = 10\%\,(v/v)$

2.5 PROBLEMS

If you have just finished reading this chapter I suggest you now try the 'pure mathematics' examples marked'*'. If you get these correct, try *all* of the applied examples 31–36. If you also get these correct, move on to the next chapter. If you do not get the right answers, re–read the explanations and try the rest of the problems 1–30, and then repeat your attempts at the applied problems.

Do each of the following calculations without, then check with, a calculator.

1. ${}^{2}/_{5} + {}^{3}/_{5}$
2. ${}^{8}/_{16} - {}^{3}/_{16}$
3.* ${}^{9}/_{15} + {}^{3}/_{7}$
4. ${}^{2}/_{3} - {}^{1}/_{6}$
5. $2\,{}^{1}/_{4} - {}^{7}/_{6}$
6. ${}^{9}/_{5} \times {}^{3}/_{7}$
7. $3\,{}^{1}/_{3} \times {}^{7}/_{15}$
8. ${}^{5}/_{6} \div {}^{3}/_{5}$
9.* $2\,{}^{1}/_{4} \div {}^{7}/_{9}$
10. ${}^{21}/_{25} \div {}^{7}/_{10}$
11. $17.4 + 13.26$
12. $17.4 - 13.26$
13. $15.01 - 13.97$
14. 7.7×3.5
15.* 9.8×17.31
16. $9.8 \div 8$
17. $13.1 \div 3.8$
18. $177.68 \div 13.5$

You can use a calculator for the following problems, but make sure you estimate the answers first.

19. $18.60 \div 3.41$
20. $18.60 \div 0.783$
21.* 3% of 21
22. 15% of 46
23. 11.2% of 31.3
24.* 15 is ?% of 45
25. 0.75 = ?%
26. $^7/_9$ = ?%
27. 81 increased by 9% gives what?
28.* What when increased by 20% gives 72?
29. 35 decreased by 20% gives?
30.* What, when decreased by 75%, gives 20?
31. The dry weight of the seeds of an annual weed is 1.75 g and of the rest of the plant 5.21 g. What percentage of the total dry weight are the seeds?
32. A typical bacterial cell is 70% water, 15% protein, and 7% nucleic acid. What weight of protein and nucleic acid is there in 3 g of bacteria? What will be the weight of matter that is not protein, nucleic acid, nor water?
33. Complete Table 2.1 which shows the relative proportions or amounts of some elements in the human body.
34. What percentage of the weight of disodium hydrogen phosphate, Na_2HPO_4, is from phosphorus?
(Use the following atomic weights of the elements: $Na = 23, H = 1, P = 31, O = 16$)
Also express this percentage as a fraction and as a decimal.

Table 2.1 Data for Question 33, Chapter 2.

Element	*Mass (kg) in 70 kg man*	*Percentage of human body weight*
Hydrogen	6.51	9.3
Carbon	13.65	
Nitrogen		5.1
Oxygen		62.8
Calcium	0.98	
Phosphorus	0.43	
Sulphur		0.64
Sodium		0.26
Potassium	0.15	
Chlorine		0.18

35. How would you make up the following sucrose or ethanol solutions?

    (i) 100 ml of a 3% (w/v) solution of sucrose.
    (ii) 500 ml of a 3% (w/v) solution of sucrose.
    (iii) 500 ml of a 5% (w/w) solution of sucrose.
    (iv) 500 ml of a 0.5% (v/v) solution of ethanol.
    (v) 20 ml of a 10% (w/v) solution of ethanol.

36. For each of the first three solutions in 35 above, calculate the weight of sucrose there would be in 5 ml.

3 RATIO AND PROPORTION: AMOUNTS, VOLUMES AND CONCENTRATIONS

3.1 INTRODUCTION TO RATIOS AND PROPORTIONS

A **ratio** is one number divided by another number but it can be written in several ways:

$$3 \div 4 = 3:4 = {}^{3}\!/\!_{4} = 3/4 = \frac{3}{4}$$

Example 3.1 If there are 20 staff and 400 students, the student : staff ratio = 400 : 20

or 400/20 or ${}^{400}\!/\!_{20}$ or $\frac{400}{20}$

= 20 : 1 or 20/1 or ${}^{20}\!/\!_{1}$ or $\frac{20}{1}$ or just 20.

Problems involving **proportions** (in a purely mathematical sense) can be represented by equations involving two ratios.

Example 3.2 If on a map, 1 cm represents 2 km, how many cm represent 12 km?

The scale of the map is a ratio of the distance on the map to the distance on the ground, and this ratio must be the same for any distance, so the problem can be represented as an equation of two ratios in which 'x' stands for the unknown number:

$$\frac{1\,\text{cm}}{2\,\text{km}} = \frac{x\,\text{cm}}{12\,\text{km}}$$

(You may find it helpful to read the first part of section 5.1 if you are unsure about solving simple equations.)
Since equations stay the same if you do the same thing to each side, to find the value of x, multiply each side by 12 km:

$$\frac{1\,\text{cm} \times 12\,\text{km}}{2\,\text{km}} = x\,\text{cm}$$

Reducing by dividing, you get

$1\,\text{cm} \times 6 = x\,\text{cm}$.

Therefore 6 cm represents 12 km.

In the example above, you could have multiplied by both denominators:

$$1\,\text{cm} \times 12\,\text{km} = x\,\text{cm} \times 2\,\text{km}$$

From this you can see that if the quantities in an equation of two ratios are represented by a, b, c and d:

$$\textbf{if } \frac{\boldsymbol{a}}{\boldsymbol{b}} = \frac{\boldsymbol{c}}{\boldsymbol{d}}$$

$$\textbf{then } \boldsymbol{a} \times \boldsymbol{d} = \boldsymbol{b} \times \boldsymbol{c}$$

Example 3.3 $\frac{x}{4} = \frac{3}{16}$

$x \times 16 = 3 \times 4$

so $16x = 12$

$x = {}^{12}/_{16} = {}^{3}/_{4}$

Example 3.4 $\frac{7}{x} = \frac{42}{36}$

$7 \times 36 = 42x$

$36 = 6x$

$x = 6$

Example 3.5 In a pharmacology experiment a piece of guinea-pig ileum is attached to a transducer so that a small contraction of the tissue produces a larger movement of a pen on a chart recorder. If a 3 mm movement of the transducer produces a 13 mm deflection on the chart recorder, what size of contraction is responsible for a 33 mm deflection on the chart recorder?

The two ratios of tissue contraction to pen movement must be the same, so if x stands for the required value,

${}^{3}/_{13} = {}^{x}/_{33}$

so $x = ({}^{3}/_{13}) \times 33\,\text{mm} = {}^{99}/_{13} = 7.6\,\text{mm}$

The most common calculations involving ratios and proportions are to do with unit conversions, and with concentrations and dilutions.

Example 3.6 1 inch = 2.54 cm

How many cm in 18 inches?

$$\frac{1\ \text{inch}}{2.54\ \text{cm}} = \frac{18\ \text{inch}}{x\ \text{cm}}$$

$$x \times 1 = 18 \times 2.54$$

$$x = 45.72\ \text{cm}$$

so 18 inches = 45.72 cm.

Example 3.7 100 ml contains 5 g of glucose. How much is there in 20 ml?

$$\frac{100\ \text{ml}}{5\ \text{g}} = \frac{20\ \text{ml}}{x\ \text{g}}$$

$$100\ \text{ml} \times x\ \text{g} = 20\ \text{ml} \times 5\ \text{g}$$

$$100x = 100\ \text{g}$$

$$x = 1\ \text{g}$$

so 20 ml contains 1 g.

Example 3.8 To 5.0 ml of a 5% solution of glucose is added water to a final volume of 250 ml. What concentration is this new solution?

You can probably do this in your head. 5.0 ml becomes 250 ml which is a 50 times bigger volume, so the concentration must be $^1/_{50} \times 5\% = 0.1\%$.

Working it as a proportion though you might be tempted to write:

$$\frac{5\ \text{ml}}{5\%} = \frac{250\ \text{ml}}{x\%}$$

This is wrong. The volume is ***not proportional*** to the concentration, but ***inversely proportional***, i.e. the larger the volume, the lower the concentration. Therefore the relationship is:

$$\frac{5\ \text{ml}}{250\ \text{ml}} = \frac{x\%}{5\%} \quad \text{or} \quad \frac{5\ \text{ml}}{x\%} = \frac{250\ \text{ml}}{5\%}$$

I think it is easier to write the relationship in the form in which you would say it:

$$5\ \text{ml of } 5\% = 250\ \text{ml of } x\%$$

For 'of' you can substitute 'times' (just as 'per' means 'divided by').

$$5\ \text{ml} \times 5\% = 250\ \text{ml} \times x\%$$

$$\text{so } x\% = \frac{5 \times 5}{250}\% = \frac{25}{250}\% = 0.1\%$$

A better way of writing out the relationship would be to include the units:

$$5\,\text{ml} \times \frac{5\,\text{g}}{100\,\text{ml}} = 250\,\text{ml} \times \frac{x\,\text{g}}{100\,\text{ml}}$$

A good reason for writing out the problem in this way is because the units might be different. For example, suppose that to 5 ml of a 5% solution is added water to a final volume of 5 litres. What concentration is the new solution?

$$5\,\text{ml} \times \frac{5\,\text{g}}{100\,\text{ml}} = 5\,\text{l} \times \frac{x\,\text{g}}{100\,\text{ml}}$$

Notice that the units do not balance, so you need to convert litres to ml. Since there are 1000 ml in 1 litre, there are 5000 ml in 5 litres.

$$5\,\text{ml} \times \frac{5\,\text{g}}{100\,\text{ml}} = 5000\,\text{ml} \times \frac{x\,\text{g}}{100\,\text{ml}}$$

$$\text{or } 5\,\text{ml} \times 5\% = 5000\,\text{ml} \times x\%$$

$$x\% = \frac{5\,\text{ml} \times 5\%}{5000\,\text{ml}}$$

$$x\% = 0.005\%$$

3.2 MOLARITY

So far I have expressed chemical concentrations only in terms of weight per volume, weight per weight, or volume per volume. A more common and usually more useful way is in terms of molarity, i.e. the moles per volume. A mole (symbol mol) is the SI unit of amount of substance.

For an element the mass of a mole of that element is its atomic mass expressed in grams instead of atomic mass units. For example, the atomic mass of carbon is 12.011, so a mole of carbon has a mass of 12.011g. It follows that **a mole of any element contains the same number of atoms as a mole of any other element.** This number, which is called Avogadro's number, is 602 000 000 000 000 000 000 000.

Similarly, for a compound, a mole is the relative molecular mass (molecular weight or formula weight) in grams. For example, 1 mole of glucose ($C_6H_{12}O_6$) contains Avogadro's number of molecules and has a mass of 180 g because its formula weight is 180. (Formula weight $= (6 \times 12) + (12 \times 1) + (6 \times 16) = 180$.)

Example 3.9 Calculate the mass of 0.25 moles of Na_2HPO_4. (Use the following atomic weights of the elements: Na = 23, H = 1, P = 31, O = 16)
Relative molecular mass (RMM) or formula weight = $(2 \times 23) + (1 \times 1) + (1 \times 31) + (4 \times 16) = 142$ g/mole
so $0.25 \text{ mol} \times 142 \text{ g/mol} = 35.5$ g.

Example 3.10 Calculate the number of moles of NaOH in 15 g.
Relative molecular mass = $23 + 16 + 1 = 40$ g/mol
So in 15 g there are $^{15\,g}/_{40\,g/mol} = 0.375$ moles.

If a solution contains 180 g glucose in 1 litre then it has a concentration of 1 mol/l. A solution containing 360 g sucrose in 1 litre also has a concentration of 1 mol/l because the relative molecular mass of sucrose is 360. The two solutions therefore have the same number of molecules per litre, but the glucose solution is 180 g/l which is the same as 18 g/100 ml or 18% (w/v) whereas the sucrose solution is 36% (w/v). In general, then, two solutions of different compounds but the same molar concentration (molarity) have the same number of molecules per litre, but usually different masses of solute per litre. A solution containing 360 g of glucose in 1 litre would have 2 moles/litre and therefore be 2 molar.

Note that the unit mol/l can also be written mol l^{-1} or M, so a 0.2 mol/l solution is the same as a 0.2 mol l^{-1} solution and a 0.2 M solution, the M being read as **molar.**

Example 3.11 What is the molarity of a solution made by dissolving 8 g of NaOH in water and diluting to 2 litres?

relative molecular mass = $23 + 16 + 1 = 40$ g/mol
number of moles = mass/RMM = $^{8\,g}/_{40\,g/mol}$
= 0.2 moles
Concentration = $^{mol}/_{l} = ^{0.2\,moles}/_{2\,litres} = 0.1$ mol/l

Example 3.12 What is the molar concentration of a 0.9%(w/v) NaCl solution?

$0.9\%(w/v) = ^{0.9\,g}/_{100\,ml} = ^{9\,g}/_{1000\,ml} = 9\,g/l$

RMM of NaCl

$= 23 + 35.5 = 58.5$ g/mol

$$\text{so molar concentration} = \frac{9\,g}{1} \times \frac{mol}{58.5\,g} = 0.15\,mol/l$$

Example 3.13 How many moles of a compound are present in 25 ml of a 0.2 mol/l solution?

Number of moles = volume × concentration
= 250 ml × 0.2 mol/1000 ml = 50/1000 = 0.05 moles

Example 3.14 What weight of glucose is needed to make 1 litre of a 0.25 M solution?

0.25 M = 0.25 mol/l
so you need 0.25 mol of glucose for 1 litre.
Since glucose has a molecular weight of 180 g/mol, weight of glucose needed
= 0.25 mol × 180 g/mol = 45 g

Example 3.15 What weight of glucose is needed to make 250 ml of a 0.5 M solution?

0.5 M = 0.5 mol/l
so you need 0.5 mol for 1 litre
0.5 mol × 180 g/mol = 90 g for 1 litre.
But you only want 250 ml, therefore you need
90 g × 250 ml/1000 ml = 22.5 g.

Rather than writing all this out, it could be combined into one expression:

$$\frac{0.5\,\text{mol}}{\text{litre}} \times \frac{180\,\text{g}}{\text{mol}} \times 250\,\text{ml} \times \frac{1\,\text{litre}}{1000\,\text{ml}} = 22.5\,\text{g}$$

Look carefully at the units. Notice that they all cancel out except for g. If you write the units in at each step of the calculation then cancel them, you can check that the calculation is correct. This is one way to make sure you do not divide by something when you should be multiplying by it. Also it is important that you report your answer with the appropriate units. **If the units are wrong, the calculation is wrong.**

3.3 DILUTIONS

It is often more practical or more convenient to make up a 'stock solution' and then dilute that to the required concentration, than to make up the required solution directly. For example, you might want to make up 1 ml of a solution of a drug (in water) at a concentration of 1 millionth of a gram per ml, but the smallest amount you can weigh accurately is 1 hundredth of a gram. It would be a bit silly to weigh out this amount and make the solution up to 10 litres (10 000 ml) to make the required concentration.

A more sensible approach is to weigh out $^1/_{100}$ g and make the solution up to 100 ml. The concentration of this 'stock solution' is

$$\frac{0.01\,\text{g}}{100\,\text{ml}} = \frac{1}{10\,000}\ \text{g/ml}$$

This is 100 times more concentrated than the solution needed. You now need to dilute a portion of this solution by a factor of 100, i.e. what is called a 100-fold dilution or 100 × dilution. A practical way of doing this would be to add 0.1 ml of the stock solution to 9.9 ml of water.

$$\frac{0.1}{9.9+0.1} = \frac{0.1}{10} = \frac{1}{100}$$

so the final solution is $^1/_{100}$ of the concentration of the stock solution.

This method uses only a single dilution step, but often it is necessary or convenient to use two or more dilution steps. This is known as 'serial dilution'.

Example 3.16 Prepare 10 ml of a 0.0001 mol/l solution (in water) of glucose from a 1 mol/l stock solution.

You can do this easily by serial dilution. Take 0.1 ml of stock + 9.9 ml of water. This is a 100-fold dilution, so the concentration of this solution (A) is 0.01 mol/l. Now take 0.1 ml of (A) and add 9.9 ml of water. Again, this is a 100-fold dilution, so the concentration of this final solution is 0.01/100 = 0.0001 mol/l.

Of course, there is no reason why dilutions should always be by factors of 10.

The fold dilution is equal to the final volume divided by the volume of the original solution. Remember that the final volume is the volume of the original solution plus the volume of the additional diluent.

Example 3.17 In a 4-fold dilution

1 ml solution + 3 ml diluent = 4 ml diluted solution

$$\text{so}\quad \frac{\text{final volume}}{\text{volume of original solution}} = \frac{4\,\text{ml}}{1\,\text{ml}}$$

Note that this means the volume of diluent you need is the final volume *minus* the volume of the original solution.

Example 3.18 To make 50 ml of a 25-fold dilution, what volumes should be used?

$$25\text{-fold dilution} = \frac{\text{final volume}}{\text{volume of original solution}} = \frac{50\,\text{ml}}{2\,\text{ml}}$$

$$\text{volume of diluent needed} = 50\,\text{ml} - 2\,\text{ml} = 48\,\text{ml}$$

Therefore add 2 ml of original solution to 48 ml diluent to get 50 ml of a 25-fold dilution.

Example 3.19 How much has a solution been diluted if 3 ml has 15 ml diluent added?

$$\text{dilution} = \frac{\text{final volume}}{\text{original volume}} = \frac{15+3}{3} = \frac{18}{3} = 6\text{-fold}$$

There are other ways of expressing dilutions. Rather than saying 'a 10-fold dilution' for example, you could say 'a 1 in 10 dilution' or 'a 1+9 dilution'. The term 'dilution factor' is usually used to mean the same as 'fold dilution'.

3.4 CONCENTRATION, VOLUME AND AMOUNT

Calculations for making up solutions, making dilutions and measuring the amount of substance produced all involve a simple relationship between concentration, volume and amount.

$$\textbf{concentration} = \frac{\textbf{amount}}{\textbf{volume}}$$

$$\text{e.g. } 0.5\text{ M} = \frac{0.5\,\text{moles}}{1\,\text{litre}}$$

or,

$$\textbf{amount} = \textbf{concentration} \times \textbf{volume}$$

$$\text{e.g. } 0.5\,\text{mol} = 0.5\,\text{molar} \times 1\,\text{litre}$$

or,

$$\textbf{volume} = \frac{\textbf{amount}}{\textbf{concentration}}$$

$$\text{e.g. } 1 \text{ litre} = {}^{0.5\,\text{mol}}/_{0.5\,\text{molar}}$$

It is really important that you distinguish between amount and concentration. For example, if you start with 1 mole of glucose and dissolve it in water to make 1 litre, the concentration is 1 mol/l (or 1M), the amount is 1 mol, and the volume is 1 litre. If you then remove 500 ml, the volume is 500 ml, the concentration is still 1 mol/l, but the amount present is 0.5 mol. If you then add 500 ml of water, the volume becomes 1 litre, the amount stays at 0.5 mol, but the concentration falls to 0.5 mol/l. (See Fig. 3.1)

Example 3.20 It was found that the concentration of NADH in a 2.0 ml sample taken from a flask containing 250 ml was 0.001 mol/l. How much NADH was in the flask?

NADH concentration = 0.001 mol/l
Total volume at this concentration = 250 ml = 0.25 l

Therefore the total amount of NADH = 0.001 mol/l × 0.25 l = 0.00025 mol.

Example 3.21 A 10 ml sample was removed from a flask containing 500 ml of an aqueous solution of X. To 1.0 ml of this sample was added 4.0 ml water, and then 2.0 ml of this final solution was put in a cuvette to measure the concentration of X. This was found to be 0.03 mol/l.

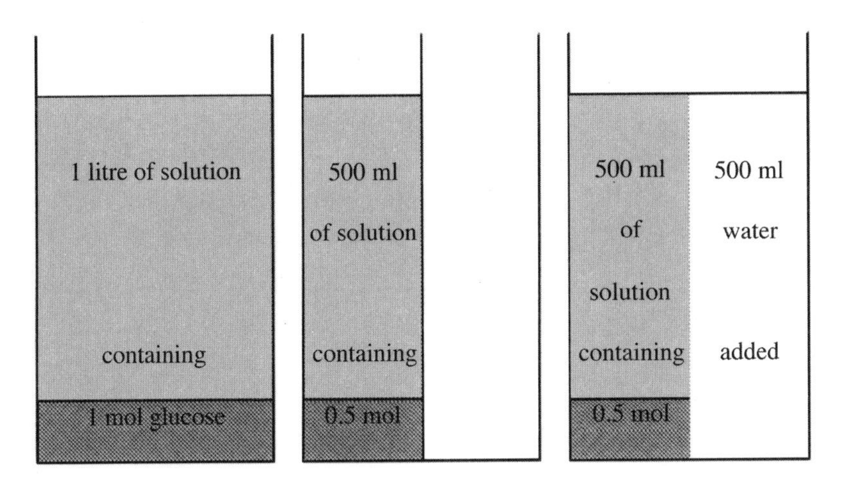

FIGURE 3.1 Relationship Between Amount, Volume and Concentration.

How much X was in the flask?

The concentration of X in the cuvette = 0.03 mol/l.
This has been diluted 5 fold,
so the original concentration of X = 0.03 mol/l × 5 = 0.15 mol/l.
Total amount of X in the flask = 0.15 mol/l × 500 ml
= 0.15 mol/l × 0.5 l
= 0.075 mol.

Example 3.22 Using the previous example, complete Table 3.1 showing the volume, amount and concentration of each of the solutions.

You can find the missing information by working down the table.

Table 3.1 Data for Example 3.22.

Volume	*Amount*	*Concentration*
2 ml	A	0.03 mol/l
5 ml	B	0.03 mol/l
1 ml	C	D
10 ml	E	F
500 ml	G	H

$$\text{Amount A} = \frac{0.03\,\text{mol}}{\text{l}} \times 2\,\text{ml} = \frac{0.03\,\text{mol}}{\text{l}} \times 0.002\ \text{l} = 0.00006\ \text{mol}$$

$$\text{Amount B} = \frac{0.03\,\text{mol}}{\text{l}} \times 5\,\text{ml} = \frac{0.03\,\text{mol}}{\text{l}} \times 0.005\ \text{l} = 0.00015\,\text{mol}$$

Amount C must be the same as B = 0.00015 mol

$$\text{Concentration D} = \frac{0.03\ \text{mol}}{\text{l}} \times \frac{(4+1)\ \text{ml}}{1\ \text{ml}} = \frac{0.15\ \text{mol}}{\text{l}}$$

$$\text{Amount E} = \frac{0.15\,\text{mol}}{\text{l}} \times 10\,\text{ml} = \frac{0.15\,\text{mol}}{\text{l}} \times 0.01\ \text{l} = 0.0015\,\text{mol}$$

Concentration F must be the same as D.

$$\text{Amount G} = \frac{0.15\,\text{mol}}{\text{l}} \times 500\,\text{ml} = \frac{0.15\,\text{mol}}{\text{l}} \times 0.5\text{l} = 0.075\,\text{mol}$$

Concentration H must be the same as D.

3.5 CONCENTRATIONS AND DILUTIONS IN PRACTICE

Often you will be interested in measuring the effects of different concentrations of a substance on some system, so you need to work out how to arrange this. Suppose the system you are using (e.g. a chemical reaction) requires that you always have a 2.0 ml volume in the tube in which the system occurs. This means that you need to work out how to achieve a range of concentrations of your substance in the 2.0 ml volume, bearing in mind first that if you start with 2.0 ml and add a volume of your test substance the system volume will increase, and second, that adding a volume of your test substance to a volume of another solution will reduce the concentration of the test substance.

Look back at the example in Section 1.2 where you have a system volume of 2.0 ml. The highest concentration of PEP you want to use is 0.0002 mol/l and you need 9 different concentrations of PEP covering a 100-fold range. The first thing to do is to think about the volumes of solution it is practical to pipette accurately. You can assume that you could use pipettes to cover a range of volumes from 10.0 ml to 0.01 ml. There are two main ways of proceeding.

(1) Preparing a Series of Dilutions

(a) Decide what is a convenient volume to pipette, e.g. 0.1 ml, and make up a series of different concentrations of PEP so that you always add a fixed volume (0.1 ml) of various concentrations to the system. You need a 1.9 ml volume of the other components of the system so that when you add 0.1 ml, you get a final volume of 2.0 ml. Since you are adding 0.1 ml of PEP and ending up with 2.0 ml, the concentration of PEP is being reduced by a factor of 20. Therefore the concentration of PEP that you make up must be $20 \times$ the concentration that you want to achieve in the system. Therefore the highest concentration of PEP that you make up must be $0.0002\,\text{mol/l} \times 20 = 0.004\,\text{mol/l}$ and the lowest must be a 100-fold dilution of this. You could make up a stock solution of PEP that is 0.004 mol/l, and from it make your series of dilutions.

Example 3.23 From a stock solution of 0.004 mol/l, a series of 9 concentrations can be made as in Table 3.2.

(b) Alternatively you could make a series of different PEP concentrations by the method of 'doubling dilutions'. Add 0.5 ml water to each tube except the first, which contains the 0.004 mol/l stock solution of PEP. Remove 0.5 ml from this tube and pipette into the second, and mix. As there is now 1.0 ml

Table 3.2 Preparing a Series of Dilutions.

Tube number	*Volume of 0.004 mol/l PEP (ml)*	*Volume of water (ml)*	*Concentration of PEP (mol/l) in tube*
1	1.00	0	0.004
2	0.75	0.25	0.003
3	0.50	0.50	0.002
4	0.25	0.75	0.001
5	0.10	0.90	0.0004
6	0.075	0.925	0.0003
7	0.050	0.950	0.0002
8	0.025	0.975	0.0001
9	0.010	0.990	0.00004
10	0	1.0	0
			Note that the concentration in the system is 1/20 of the above values.

of solution in this second tube, the concentration will have been halved, being now 0.002 mol/l. Pipette 0.5 ml from this tube into the third tube. This again halves the concentration so producing a 0.001 mol/l solution. Continue in this way for the whole set of tubes. You will end up with a series of tubes with each concentration being one half of the previous one: 0.004, 0.002, 0.001, 0.0005, 0.00025, 0.000125, 0.0000625, 0.00003125, and 0.000015625 mol/l. Notice that it only requires 8 tubes to cover the 100-fold range, but the numbers would be rather awkward to handle if you were going to plot a graph of the results.

(2) Using Different Volumes of a Stock Solution

A different approach is to pipette different volumes of your stock solution, plus different volumes of water, directly into the system. This time the concentration of the stock solution you make up needs to be lower because you will start by pipetting a larger volume, and the volume of the other components of the system must be smaller.

Example 3.24 Using a stock solution of PEP that is 0.004 mol/l, the required system concentrations can be made by adding stock and water to 1 ml of the other components as in Table 3.3.

Table 3.3 Using Different Volumes of a Stock Solution to Prepare Dilutions.

Volume of stock 0.0004 mol/l PEP (ml)	*Volume of water (ml)*	*Dilution factor (Remember the final volume is 2.0 ml)*	*Concentration in the system (mol/l)*
1.0	0	2	0.00020
0.75	0.25	8/3	0.00015
0.50	0.50	4	0.00010
0.25	0.75	8	0.000050
0.10	0.90	20	0.000020
0.075	0.925	80/3	0.000015
0.050	0.950	40	0.000010
0.025	0.975	80	0.000005
0.010	0.990	200	0.000002
0	1.0		0

There are advantages and disadvantages with each method. Using the method (1a) involves three pipettings for each concentration, and the use of more tubes. Method (2) involves only two pipettings (so should be more accurate) and no tubes, but if you end up pipetting very small volumes accuracy may suffer. The method of doubling dilutions is attractive because you do not need to think about changing the volumes you are pipetting, but it involves three pipettings for each concentration, and any error in one tube will be carried on to affect subsequent ones.

In microbiology or other cell culture experiments, serial dilution by a factor of 10 is often used to dilute a cell culture enough so that the number of cells per ml can be counted, or a dilute culture spread on a plate so that each cell grows to form a separate colony. A typical procedure would be to set up a series of tubes containing 9 ml of medium, and to add 1ml of the original culture to the first, mix, and pipette 1ml of this into the next tube, and so on. The number of cells per ml in each tube will then be $^1/_{10}$ of that in the previous tube. In this way it is possible to prepare quickly a wide range of concentrations.

3.6. PROBLEMS

If you have just finished reading this Chapter I suggest you now try the 'pure mathematics' examples marked'*'. If you get these correct, try *all* of the applied examples. If you also get these correct, move on to the next chapter.

If you do not get the right answers, re–read the explanations and try the rest of the problems 1–12, and then repeat your attempts at the applied problems.

Write the following ratios as fractions in their simplest forms.

1. 12 : 6	2. 24 : 15	**3.*** 9 : 72
4. 13 : 52	5. 36 g : 84 g	**6.*** 210 ml : 71

Solve for x.

7. $x/9 = 4/3$	8. $2/9 = x/3$	**9.*** $11/8 = 44/x$
10. $56/x = 8/7$	11. $1/x = 0.5/1$	**12.*** $5/x = 3/0.5$

13. If a map is drawn 1:25 000, what distance does 2 cm represent?

14.* What scale is a map where 1 cm represents 5 km?

15. How many litres in 1 gallon if 1 pint = 567.5 ml and there are 8 pints in 1 gallon?

16. Ecologists sometimes use the capture/recapture technique to estimate the population of an animal in a certain study area. For example, for voles the method consists of trapping a sample of voles in traps scattered throughout the area. The trapped voles are marked in a way which enables them to be identified but does not affect them. The marked voles are released and the traps reset a day later. The total number of voles captured this time, and the number recaptured which are marked, can be used to estimate the total population. The proportion of marked voles recaptured to total marked voles is the same as the proportion of total voles captured to the total population. (Note that this assumes there is no change in the habits, life-expectancy, etc. of the marked voles.)

    If initially 50 voles are captured, marked, and released, and later 40 voles are captured, of which 5 carry the mark, what is the total vole population in the area surveyed?

17. Given a 5% (w/v) solution of sodium chloride (formula weight = 58.5 g) complete Table 3.4.

18. The formula weight of ATP (disodium salt) is 551.

    (i) What is the weight of 0.5 mol of this?
    (ii) What weight of this ATP is required for 1 litre of a 0.5 mol/l solution?
    (iii) What weight of this ATP is required for 200 ml of a 0.25 mol/l solution?
    (iv) What is the % (w/v) concentration of a 0.5 mol/l solution of ATP?
    (v) What is the % (w/v) concentration of a solution of ATP containing 0.0187 g in 50 ml?

Table 3.4 Data for Question 17, Chapter 3.

Volume of 5% (w/v) solution	*Mass in given volume*	*Moles of Na^+ in given volume*
100 ml		
1 l		
5 ml		
0.01 ml		
20 l		

(vi) What is the molar concentration of a solution of ATP containing 0.0374 g in 30 ml?
(vii) What volume of 0.01 mol/l ATP solution could you make with 1 g of ATP?
(viii) What volume of 0.005 mol/l ATP solution could you make from 0.2 ml of a 1% (w/v) solution?
(ix) Given 100 ml of 0.01 mol/l ATP solution, how would you make up a set of solutions of 10 ml volume each, with concentrations of 0.005, 0.004, 0.003, 0.002, 0.001, and 0 mol/l?
(x) If 0.05 ml of each of the six solutions in (ix) above were pipetted into a cuvette containing 1.95 ml of other solutions, what would be the final ATP concentration in each cuvette?

19. Given 100 ml of a 0.5 mol/l solution of glucose (formula weight 180), how would you prepare the following? (Assume the smallest volume you can measure accurately is 0.05 ml and that you need to be reasonably economical with the stock solution.)

(i) 10 ml of a 0.02 mol/l solution.
(ii) 10 ml of a 0.004 mol/l solution.
(iii) 1.0 ml of a 0.01 mol/l solution.
(iv) 1.0 ml of a 0.03 mol/l solution.
(v) 10 ml of a 0.1% (w/v) solution.
(vi) 5.0 ml of a 0.03% (w/v) solution.
(vii) 2.0 ml of a 0.02% (w/v) solution.
(viii) 1.0 ml of a 0.15% (w/v) solution.
(ix) A set of 5 tubes containing 1.0 ml of glucose solutions with the concentrations 0.02, 0.015, 0.01, 0.005, and 0 mol/l.
(x) A set of 5 tubes containing 10 ml of glucose solutions with the concentrations 0.20, 0.15, 0.10, 0.05, and 0% (w/v).

20. Inulin (not to be confused with insulin) is a soluble polysaccharide that if injected into a vein, is filtered from the blood plasma by the

glomeruli of the kidneys in the same concentration as in plasma. It is neither reabsorbed nor excreted in the tubules, and so appears in the urine. Therefore its rate of disappearance from the plasma is a measure of the 'glomerular filtration rate' (GFR, the volume filtered by the glomeruli per minute) as is its rate of appearance in the urine. Measurement of GFR can give a useful indication of kidney function.

Calculate the GFR in ml/minute if the rate of inulin excretion into the urine is 120 mg/minute when the plasma concentration is constant at $^{100\text{ mg}}/_{100\text{ml}}$.

4 EXPONENTS AND PREFIXES: SCIENTIFIC NOTATION, CONVERSION OF UNITS, AND MORE ON CONCENTRATIONS

4.1 INTRODUCTION

You may have noticed that I have so far avoided using many multiples of units. For example, for volumes I have only used litres or millilitres, and for weights, grams or kilograms. This was because I assumed you would be familiar with these units and could confidently convert from one to the other, but may not be so competent at using others. However, you may also have noticed that in restricting myself I have had to write some cumbersome numbers, e.g. 2 312 000, 10 000 ml, and 0.0001 mol/l. To avoid writing numbers like this I could have expressed them either in scientific notation or by using a different unit. To use either you need to understand exponents.

4.2 EXPONENTS AND SCIENTIFIC NOTATION

You can express many numbers as being the same as other numbers multiplied together, i.e. as the product of factors.

Example 4.1 $10 = 5 \times 2$, or $12 = 4 \times 3$
5 and 2 are factors of 10. 4 and 3 are factors of 12.
For some numbers, the factors can be the same.

Example 4.2

$100 = 10 \times 10$	$4 = 2 \times 2$
$9 = 3 \times 3$	$1000 = 10 \times 10 \times 10$
$16 = 2 \times 2 \times 2 \times 2$	$81 = 3 \times 3 \times 3 \times 3$

Shorthand ways of writing these examples are:

$10 \times 10 = 10^2$	$2 \times 2 = 2^2$
$3 \times 3 = 3^2$	$10 \times 10 \times 10 = 10^3$
$2 \times 2 \times 2 \times 2 = 2^4$	$3 \times 3 \times 3 \times 3 = 3^4$

Notice that in each case, the shorthand version consists of the number that is being multiplied by itself, which is known as the **base**, and a raised number representing how many times the base is written down, which is called the **exponent** or **power**. (Powers of 2 are usually read as, for example, '10 squared' and powers of 3 as '10 cubed'. 2^4 can be read as '2 to the 4th power' or '2 to the power 4' or more simply '2 to the 4'.)

Example 4.3 For the number 2^4, the base is 2 and the exponent or power is 4.

For any non-zero number a and any non-zero whole number b,

$$\boldsymbol{a^b = a \times a \times a \times a....} \quad \text{where } a \text{ is written } b \text{ times.}$$

To **multiply exponential numbers having the same base**, add the exponents. For any non-zero base a and any non-zero whole numbers b and c,

$$\boldsymbol{a^b \times a^c = a^{b+c}}$$

Example 4.4 $10^2 \times 10^3 = ?$

$10^2 \times 10^3 = 10^{2+3} = 10^5$
This is the same as $100 \times 1000 = 100\ 000$

Example 4.5 $2^2 \times 2^3 = ?$

$2^2 \times 2^3 = 2^5$
This is the same as $4 \times 8 = 32$

Note that to do this, the bases must be the same. If they are not, it may be possible to make them the same.

Example 4.6 $4^2 \times 2^4 = ?$

Since 4 is $2 \times 2, 4^2$ can be written as
$2 \times 2 \times 2 \times 2$ which $= 2^4$
So $4^2 \times 2^4 = 2^4 \times 2^4 = 2^8$

To **divide exponential numbers with the same base**, subtract the exponents.

$$\boldsymbol{a^b / a^c = a^{b-c}}$$

Example 4.7 $10^3/10^2 = ?$

$10^3/10^2 = 10^{3-2} = 10^1 = 10$

Example 4.8 $2^3/2^4 = ?$

$2^3/2^4 = 2^{3-4} = 2^{-1}$

What does 2^{-1} mean?
For any non-zero base a and any non-zero whole number b,

$$\boldsymbol{a^{-b} = 1/a^b}$$

Example 4.9 What is 10^{-3} as a fraction?

$10^{-3} = 1/10^3$ which $= {}^1/_{1000}$

Example 4.10 What is 3^{-4}?

$3^{-4} = 1/3^4 = {}^1/_{81}$

Example 4.11 Simplify $3^3/3^3$

One number divided by itself obviously equals 1.
Therefore any exponential number divided by itself is one.
So, $3^3/3^3 = 1$
But also $3^3/3^3 = 3^{3-3} = 3^0$
So $3^0 = 1$

In fact, **for any non-zero base a, a^0** $= 1$.

Example 4.12 $5^0 = 10^0 = 2^0 = 3^0 = 1$

It is possible that you may need to simplify a number that is a power of an exponential number.

Example 4.13 $(2^3)^2 = ?$

$(2^3)^2 = 2^3 \times 2^3 = 2^{3+3} = 2^6$

Example 4.14 $(4^3)^4 = ?$

$(4^3)^4 = 4^3 \times 4^3 \times 4^3 \times 4^3$
$= 4^{3+3+3+3} = 4^{12}$
In general then, $(a^b)^c = a^{b \times c}$

If you write out values of the powers of 10 you can see that for each increase of 1 in the power, the value increases by a factor of 10. This is shown in Table 4.1.

Table 4.1 Powers and Values.

Power	−3	−2	−1	0	1	2	3	4
Number	10^{-3}	10^{-2}	10^{-1}	10^0	10^1	10^2	10^3	10^4
Value	${}^1/_{1000}$	${}^1/_{100}$	${}^1/_{10}$	1	10	100	1000	10 000
Decimal	0.001	0.01	0.1	1	10	100	1000	10 000

Multiplying or dividing by 10^b is therefore the same as moving the decimal point b places to the right or left respectively.

Example 4.15 $3.0 \times 10^3 = 3000$
decimal point moved 3 places to the right
$3.0 \times 10^{-3} = 0.003$
decimal point moved 3 places to the left
$0.003 \times 10^4 = 30$
decimal point moved 4 places to the right

Example 4.16 $0.003/10^{-4} = ?$

$10^{-4} = 1/10^4$
Since dividing by a fraction is the same as multiplying by the inverse of the fraction,
$0.003/10^{-4} = 0.003 \times 10^4 = 30$

Example 4.17 $38.1 = 3810 \times 10^{-2}$ or 381.0×10^{-1}
or 3.81×10^1 or 0.381×10^2

Exponents become particularly useful when the base is 10 because it allows us to write out large numbers in a concise way.

Example 4.18 Avogadro's number = 602 200 000 000 000 000 000 000. This can be written as 6.022×10^{23} because it is the same as $6.022 \times 10 \times 10 \times 10 \times 10$ ...... (23 times).

In general, numbers written in the form $a \times 10^b$ where a is greater than 1 but less than 10 and b is a non-zero whole number are said to be expressed in **scientific notation**.

To **express a number in scientific notation**, find 'a' by moving the decimal point to leave one non-zero digit to the left of the decimal point. Calculate the value of the exponent by counting the number of places you have moved the decimal point. If you have moved the decimal point to the left, b is positive, and if to the right, negative.

Example 4.19 $24\,000 = 2.4 \times 10^4$
because the decimal point is moved 4 places to the left.

Example 4.20 $0.000\,024 = 2.4 \times 10^{-5}$
because the decimal point is moved 5 places to the right.

To add or subtract numbers in scientific notation, you need to express them in the same power.

Example 4.21 $3.7 \times 10^4 - 2.4 \times 10^3 = ?$

$3.7 \times 10^4 = 37 \times 10^3$
So $3.7 \times 10^4 - 2.4 \times 10^3 = 37 \times 10^3 - 2.4 \times 10^3$
$34.6 \times 10^3 = 3.46 \times 10^4$

To multiply or divide numbers in scientific notation, deal with the two parts separately.

Example 4.22 $(2 \times 10^4) \times (6 \times 10^3) = ?$

$(2 \times 6) \times (10^4 \times 10^3) = 12 \times 10^7 = 1.2 \times 10^8$

Example 4.23 $\dfrac{3.36 \times 10^3}{2 \times 10^4} = \dfrac{3.36}{2} \times \dfrac{10^3}{10^4} = 1.68 \times 10^{-1}$

How you express the *final* value is a matter of choice. 1.68×10^{-1} is more concisely written as 0.168. Many people prefer to move the decimal point one or two places either way rather than write the exponential form.

Example 4.24 320 rather than 3.2×10^2
32 rather than 3.2×10^1
0.32 rather than 3.2×10^{-1}
0.032 rather than 3.2×10^{-2}

If the value ends up having a power less than –2 or greater than +2, then you can either leave it in scientific notation, or change the units by using a prefix. This is covered in the next section. Before going on to this however, try the following example on your calculator.

Example 4.25 $3.6 \times 10^3 - 0.8 \times 10^3 = ?$

The correct procedure is to key, in order: 3.6,EXP,3,–,0.8,EXP,3, =. A common mistake is to key 3.6,×,10,EXP,3,–,0.8,×,10,EXP,3 =, which gives an answer 10 times too big. EXP on the calculator stands for exponent with base 10, so you should not key in the × 10. Similarly, if you wanted to divide by 10^4, you should key: ÷,1,EXP,4, and not ÷,10,EXP,4.

4.3 METRIC PREFIXES: CHANGING THE UNITS

You have already come across the prefixes 'milli', meaning one thousandth and 'kilo' meaning one thousand. Prefixes like these are another way to avoid cumbersome numbers.

Example 4.26 $0.0004 \text{ m} = 4 \times 10^{-4} \text{ m} = 0.4 \times 10^{-3} \text{ m} = 0.4 \text{ mm}$

Example 4.27 $400\,000 \text{ g} = 4 \times 10^5 \text{ g} = 400 \times 10^3 \text{ g} = 400 \text{ kg}$

As you can see in the examples above, converting from one unit to another involves multiplying or dividing by a power of 10.

There are many prefixes. The most commonly used are shown in Table 4.2.

Example 4.28 1 μg = 1 microgram = 1×10^{-6} g or 0.000 001 g

It is unfortunate that some of the letters used as symbols for the prefixes are the same as, or similar to, those used for other prefixes or units. In particular, confusion can arise between G (giga) and g (gram), and between M, m, and μ (mega, milli, and micro) and m and M (metre and molar). It is therefore important that you use the correct symbols for the prefix and unit, *do not invent your own abbreviations*, and preferably use only correct SI units, except where others make more sense. It is usually best also to restrict yourself to multiples of 1000, e.g. kilo, milli, and micro. Below are some examples to show the problems.

Example 4.29 'a 10 mM solution...'

Since the M follows the m, it cannot mean mega, therefore it must mean molar, and so mM probably means millimolar since a '10 metremolar solution' does not make much sense. It would be better to express this as 'a 10 mmol/l solution' or 'a 10 mmol l^{-1} solution, or for strict SI enthusiasts, 'a 10 mol m^{-3} solution'.

Example 4.30 'a rate of 5 μM $l^{-1}m^{-1}$'

Table 4.2 Commonly Used Prefixes to Denote Multiples of Units.

Prefix	*Symbol*	*Multiple*
giga	G	10^9
mega	M	10^6
kilo	k	10^3
hecto	h	10^2
deca	da	10^1
base unit	–	1
deci	d	10^{-1}
centi	c	10^{-2}
milli	m	10^{-3}
micro	μ	10^{-6}
nano	n	10^{-9}
pico	p	10^{-12}
femto	f	10^{-15}

Since μM is micromolar, this appears to mean a rate of 5 μmol/l per litre per metre, a concentration per volume per length, when what is probably meant is a rate of 5 μmol/l per minute, a change in concentration with time.

Example 4.31 'a concentration of 0.5 mmol dm^{-3}...'

This is acceptable but could more simply be expressed as 0.5 mmol/l, 0.5 mmol l^{-1}, or 0.5 mol m^{-3}.

Some people say that since a litre is not a proper SI unit, volumes should be based on a unit of length to the power 3. They then use dm^3 or cm^3 or mm^3. However, this is not really in the spirit of SI either as one should not use powers of multiples of units. You do need to be able to convert between these units though.

Example 4.32 To convert cm^3 to m^3

$$1\,\text{cm} = 10^{-2}\,\text{m}$$
$$\text{so } (1\,\text{cm})^3 = (10^{-2}\,\text{m})^3 = 10^{-6}\,\text{m}^3$$

Example 4.33 To convert cm^3 to litres

$$1\,\text{cm} = 10^{-1}\,\text{dm}$$
$$\text{so } (1\,\text{cm})^3 = (10^{-1}\,\text{dm})^3 = 10^{-3}\,\text{dm}^3 = 10^{-3}\,\text{l} = 1\,\text{ml}.$$

Example 4.34 What is the equivalent in litres of 1 mm^3?

$$1\,\text{mm}^3 = (10^{-3}\,\text{m})^3 = 10^{-9}\,\text{m}^3$$
$$= 10^{-6}\,\text{dm}^3 = 10^{-6}\,\text{l} = 1\,\mu\text{l}.$$

Concentrations in cells are typically millimolar or micromolar for metabolites of central pathways, and pico or femtomolar for messengers. Some equivalent units are:

$$1\,\text{mol/m}^3 = 1\,\text{mmol/l} = 1\,\mu\text{mol/ml} = 1\,\text{nmol}/\mu\text{l}$$
$$1\,\text{mmol/m}^3 = 1\,\mu\text{mol/l} = 1\,\text{nmol/ml} = 1\,\text{pmol}/\mu\text{l}\,.$$
$$1\,\mu\text{mol/m}^3 = 1\,\text{nmol/l} = 1\,\text{pmol/ml} = 1\,\text{fmol}/\mu\text{l}$$

4.4 PROBLEMS

If you have just finished reading this chapter I suggest you now try the 'pure mathematics' examples marked'*'. If you get these correct, try *all* of the applied examples. If you also get these correct, move on to the next chapter. If you do not get the right answers, re–read the explanations and try the rest of the problems 1–59, and then repeat your attempts at the applied problems.

Evaluate the expressions in questions 1–21. **Use a calculator only to check your answers.**

1. 2^3 **2.*** 5^0 3. 3^{-3}
4. $(0.1)^2$ 5. $(0.3)^4$ **6.*** $(0.5)^{-1}$
7. $(2/3)^2$ **8.*** $(3/2)^{-2}$ 9. $(4/5)^{-4}$
10. $a^2 \times a^3$ 11. $a^{-2} \times a^3$ **12.*** $a^{-3} \times a^{-2}$
13. a^5/a^3 14. a^{-2}/a^3 15. a^{-3}/a^{-2}
16. $(a^2)^3$ **17.*** $(a^2)^{-1}$ 18. $(a^{-2})^{-3}$
19. $10^3/10^2$ **20.*** 100×10^{-3} 21. $10^2/1000$

Express the following in scientific notation.

22. 0.000035 23. 17 300 000 000 **24.*** 3184.576
25. 373×10^{-6} 26. 0.016×10^4 **27.*** 0.0351×10^{-2}

Express the following as decimals.

28. 3.15×10^{-4} 29. 37.6×10^{-5} **30.*** 0.015×10^4
31. 1.5×10^{-2} **32.*** 478.6×10^{-4} 33. 0.0061×10^4

Evaluate the following (questions 34–45) using a calculator only to check the answers.

34. $4.51 \times 10^3 + 6.42 \times 10^3$ **35.*** $1.32 \times 10^8 + 8.68 \times 10^8$
36. $4.51 \times 10^4 - 1.62 \times 10^4$ **37.*** $1.32 \times 10^{-2} - 4.36 \times 10^{-2}$
38. $4.51 \times 10^{-2} - 0.15 \times 10^{-3}$ 39. $1.5 \times 10^{-3} - 4.5 \times 10^{-2}$
40. $(4.5 \times 10^3)(6.0 \times 10^{-2})$ **41.*** $4.5 \times 10^3/6.0 \times 10^{-2}$
42. $3.6 \times 10^{-5} \times 11 \times 10^2$ 43. $3.6 \times 10^{-5}/12 \times 10^2$
44. $3.6 \times 10^{-3}/12 \times 10^{-2}$ **45.*** $21 \times 10^3 \times 5 \times 10^{-6}/35 \times 10^{-2}$

Convert into the required units.

46. 471 mg = ? g 47. 18 ml = ? l
48.* 14×10^{-5} l = ? μl 49. 0.000051 g = ? μg
50. 42 μg = ? mg 51. 831 μl = ? ml

52.* 0.14×10^{-2} ml = ? µl
53. 0.037 mmol/l = ? µmol/l
54. 3.1×10^{-5}mol/l = ? mmol/l
55.* 17 mmol/l = ? mol/m^3
56. 0.031×10^5 nmol/l = ? µmol/l
57.* 14 pmol/µl = ? mmol/ml
58. 0.15% (w/v) = ? mg/l
59.* 3.1×10^{-4} mg/ml = ? % (w/v)

60. Complete Table 4.3, which shows the composition of phosphate-buffered saline. Use a calculator if necessary. Use the following atomic weights: Na = 23, K = 39, P = 31, O = 16, H = 1, Cl = 35.5, Mg = 24, Ca = 40

Table 4.3 Data for Question 60, Chapter 4.

Components	*Formula weight*	*Concentration % (w/v)*	*Concentration mmol/l*
Na_2HPO_4	142	0.115	
KH_2PO_4	136		1.5
NaCl	58.5	0.80	
KCl	74.5		2.68
$MgCl_2 \cdot 6H_2O$	203		0.49
$CaCl_2 \cdot 2H_2O$	147	0.013	

61. What are the total phosphate and total chloride concentrations in this solution?
Express your answers as both %(w/v) and mmol/l.
62. What are the concentrations of each of the metal ions in this solution? Express your answers as both %(w/v) and mmol/l.
63. Sodium pyruvate has a formula weight of 110. Calculate the weight required to make up 100 ml of a 50 mmol/l solution. Complete Table 4.4

Table 4.4 Data for Question 63, Chapter 4.

Volume of 50 mmol/l solution	*Volume of water added*	*Concentration*	*Weight of sodium pyruvate in 0.2 ml of this solution*
10 ml	90 ml		
0.5 ml	19.5 ml		
100 µl	0.9 ml		
25 µl	475 µl		
1.5×10^{-2} l	3.5×10^4 ml		
	20 ml		0.55 mg
	490 µl		2.2×10^{-5} g
70 µl			7.7×10^{-4} mg

which shows the composition of various solutions made from this stock solution.

64. A bacterial cell culture has 1.0 ml removed and added to 9.0 ml of medium. 0.1 ml of this is removed and added to 9.9 ml of medium. 1 ml of this is removed and added to 9 ml of medium. 1 ml of this is removed and added to 9 ml of medium. The number of bacteria in 0.1 ml of the final dilution was found to be 23. What was the approximate number of bacteria in the original culture?

5 SOLVING EQUATIONS AND EVALUATING EXPRESSIONS

5.1 SOLVING EQUATIONS

You have already come across equations of two ratios in Chapter 3. An algebraic equation will usually show the relationship between a variable represented by a letter such as x or y and something else. To solve the equation means to find the value(s) of x that make(s) the equation true. Different equations that have the same solution are called **equivalent equations**.

Example 5.1 $5y + 6 = 36$
is equivalent to $5y = 30$
and the solution is $y = 6$
because when $y = 6$,
$(5 \times 6) + 6 = 36$ is true.

In order to solve equations it is necessary to change them into simpler equivalent equations. You can do this by adding to or subtracting from both sides of the equation the same non-zero number. You can multiply or divide both sides of the equation by the same non-zero number. You may also substitute another expression for part of the original equation.

Example 5.2 $5y + 6 = 36$
Subtracting 6 from both sides gives
$5y + 6 - 6 = 36 - 6$
so $5y = 30$
Dividing both sides by 5,
$5y/5 = 30/5$
so $y = 6$

Example 5.3 $5x/6 = 5$

Dividing both sides by 5 gives

$$\frac{5x}{6 \times 5} = \frac{5}{5}$$

so $\frac{x}{6} = 1$

Multiplying both sides by 6 gives

$$\frac{x \times 6}{6} = 1 \times 6$$

so $x = 6$

Example 5.4 $7x + 2 - 2x + 4 = 36$

Combining the x terms,

$7x - 2x = 5x$

Combining the numbers,

$2 + 4 = 6$

So $5x + 6 = 36$

So $x = 6$

Example 5.5 $4 + 9y - 15y = 1$

Combining the y terms

$4 - 6y = 1$

Subtracting 4 from both sides

$-6y = -3$

Dividing both sides by -6

$y = \frac{1}{2}$

Example 5.6 $5(x + 3) = 3x - 4$

Expanding the left hand side

$5x + 15 = 3x - 4$

Subtracting $3x$ from both sides

$2x + 15 = -4$

Subtracting 15 from both sides

$2x = -19$

$x = -9.5$

Note that when **multiplying or dividing numbers in parentheses** (brackets) you **must** multiply or divide **all** the numbers.

Example 5.7 $(x + 3)(x - 3) = 16$

Expanding the left hand side

$x^2 + 3x - 3x - 9 = 16$

Combining the x terms

$x^2 - 9 = 16$

Adding 9 to both sides

$x^2 = 25$

and because $25 = 5 \times 5$,
$x = 5$

As I showed at the beginning of Chapter 3, if the numerators and the denominators of two ratios are represented by letters,

$$\textbf{if} \quad \frac{a}{b} = \frac{c}{d}$$

$$\textbf{then} \quad a \times d = b \times c$$

This gives a quicker way of simplifying proportions than multiplying both sides of the equation by both denominators.

Example 5.8 $\dfrac{5}{2x} = \dfrac{3}{4}$

Multiplying 5 by 4, and 3 by $2x$ gives

$20 = 6x$

Dividing by 6

$x = {}^{20}/_{6} = 3.33$

Example 5.9 $\dfrac{x-5}{x+2} = \dfrac{2}{9}$

$9(x-5) = 2(x+2)$

Expanding each side

$9x - 45 = 2x + 4$

Subtracting $2x$ from both sides

$7x - 45 = 4$

Adding 45 to both sides

$7x = 49$

Dividing both sides by 7

$x = 7$

The solution of an equation of this kind is not always so simple: sometimes there are two possible values for x.

Example 5.10 $\dfrac{x+7}{x+1} = \dfrac{6}{x}$

Multiplying $(x+7)$ by x, and 6 by $(x+1)$

$x^2 + 7x = 6x + 6$

Subtracting $6x + 6$ from both sides

$x^2 + x - 6 = 0$

We now need a way of finding x. One way of doing this is to express the equation as two factors.

$x^2 + x - 6 = 0$

is the same as

$(x + 3)(x - 2) = 0$

For this to be true, one of the factors must $= 0$

so either

$x + 3 = 0$ or $x - 2 = 0$

$x = -3$ or $x = 2$

To check these two solutions,

substitute into the original equation

If $x = -3$, $\frac{-3 + 7}{-3 + 1} = \frac{6}{-3}$

$\frac{4}{-2} = \frac{6}{-3}$ or $-2 = -2$

If $x = 2$, $\frac{2 + 7}{2 + 1} = \frac{6}{2}$

$\frac{9}{3} = \frac{6}{2}$ or $3 = 3$

so both of these values are valid solutions.

Equations where the unknown variable x is present to the power 2 (but not greater than 2) are known as **quadratic equations**. The last example was therefore finding the solution of a quadratic equation. In this example, it was possible to see how it could be factored into the form $a \times b = 0$, and then, in turn a and b are set equal to 0.

However, many quadratic equations cannot be easily factored and another method using a formula has to be used. For an equation of the type $ax^2 + bx + c = 0$, where a is a non-zero number, the solutions are given by:

$$\boldsymbol{x = \frac{-b \pm \sqrt{b^2 - 4ac}}{2a}}$$

$\sqrt{\ }$ means' the square root of what is under the sign'.

A square root of a number is one of its equal factors.

$\sqrt{b} =$ the number a such that $a^2 = b$

It is not essential that you know how this formula is derived. However, you should be aware of its existence and be able to use it.

Example 5.11 Solve the equation $x^2 - 5x = 6$

Set the equation to $= 0$

$x^2 - 5x - 6 = 0$

Substitute the appropriate values in the formula

$a = 1, b = -5, c = -6$

$$\text{so } x = \frac{-(-5) \pm \sqrt{(-5)^2 - 4(1)(-6)}}{2(1)}$$

$$\text{so } x = \frac{5 \pm \sqrt{49}}{2} = \frac{5 \pm 7}{2} = 6 \text{ or } -1$$

Note these are the same solutions as you would get by factoring.

$x^2 - 5x - 6 = 0$
so $(x - 6)(x + 1) = 0$
so $x = 6$ or $x = -1$

5.2 EVALUATING EXPRESSIONS

To evaluate an expression, substitute the given values for the unknown variable. You should already be familiar with this for simple examples.

Example 5.12 The circumference c of a circle is given by the formula $c = \pi d$ where d is the diameter. What is the circumference of a circle with a diameter of 5 m?

$c = \pi(5) = 15.7\,\text{m}$

Example 5.13 What is the radius of a circle with a circumference of 15.7 m?

$c = \pi d$ rearranges to $d = c/\pi$
$d = 15.7/\pi = 5$
radius $= d/2 = 5/2 = 2.5\,\text{m}$

Example 5.14 What is the area A of a circle of radius 3 m if the area is given by the formula $A = \pi r^2$ where r is the radius?

$A = \pi(3)^2 = 28.3\,\text{m}^2$

Example 5.15 The lengths of the three sides of a right-angled triangle are related by the equation $x^2 + y^2 = z^2$ where x, y, and z are the lengths of the sides, z being the side opposite the right angle. What is the length of side x when $z = 5$ m and $y = 4$ m?

$x^2 + 4^2 = 5^2$
$x^2 + 16 = 25$
$x^2 = 25 - 16 = 9$
so length of side $x = 3$ m.

It is important when evaluating expressions that you are careful that you substitute values with the correct units.

Example 5.16 If the amount of calcium in a 1.0 g sample of bread is 5.0 mg, what is the calcium concentration expressed as %(w/w)?

%(w/w) = (weight of calcium/weight of bread) × 100.
Direct substitution would give
calcium concentration = (5/1) × 100 = 500

This is wrong because the units do not match. The correct calculation is

$$\frac{5.0\,\text{mg}}{1.0\,\text{g}} \times \frac{100}{1} \times \frac{1\,\text{g}}{1000\,\text{mg}} = 0.5\%$$

Example 5.17 In the formula $A = \text{E}\,c\,d$
A = the absorbance of a solution as measured in a spectrophotometer,
E = the absorption coefficient of the solute
c = the concentration of the solute
d = the length of the light path through the solution
What is the absorbance of a solution of 0.1 mmol/1 p-nitrophenol if the absorption coefficient is 1.8×10^4 1 $\text{mol}^{-1}\text{cm}^{-1}$ and a light path of 10 mm is used?

$$A = \frac{1.8 \times 10^4\,1}{\text{mol cm}} \times \frac{0.1\,\text{mmol}}{1} \times 10\,\text{mm}$$

Changing the units to match

$$A = \frac{1.8 \times 10^4\,1}{\text{mol cm}} \times \frac{0.1 \times 10^{-3}\,\text{mol}}{1} \times 10 \times 10^{-1}\,\text{cm}$$

so $A = 1.8$

Note that in this formula A has no units, so the units of E must be the inverse of the units of $c \times d$, i.e. the units of E are litres/(mol × cm) since the units of $c \times d$ are (mol/1) × cm.

Example 5.18 The equilibrium constant, K_{eq}, for a reaction $A + B \rightarrow C$ is given by $K_{eq} = \dfrac{[C]}{[A][B]}$

where $[A]$, $[B]$, and $[C]$ represent the concentrations of the reactants A and B and of the product C at equilibrium. Calculate the equilibrium constant if $A = 0.2$ mol/1, $B = 0.5$ mmol/1, and $C = 15$ μmol/l.

$$K_{eq} = \frac{15 \times 10^{-6}\ \text{mol/l}}{(0.2\ \text{mol/l})(0.5 \times 10^{-3}\ \text{mol/1})}$$

$$K_{eq} = \frac{15 \times 10^{-6}\ 1}{(0.2)(0.5)(10^{-3})\ \text{mol}}$$

$$K_{eq} = \frac{1.5 \times 10^{-5}\,1}{10^{-4}\,\text{mol}} = 0.15\ 1/\text{mol}$$

5.3 PROBLEMS

If you have just finished reading this chapter I suggest you now try the 'pure mathematics' examples marked'*'. If you get these correct, try *all* of the applied examples. If you also get these correct, move on to the next chapter. If you do not get the right answers, re–read the explanations and try the rest of the problems 1–24, and then repeat your attempts at the applied problems.

Solve each of the following equations.

1. $3x + 15 = 6$
2.* $7 - 2x = 11$
3. $11x + 5 - 3x - 7 = 42$
4. $7x - 15 = 5 - 3x$
5. $7(2x - 3) = 21$
6.* $4(11 - 3x) = 2x + 16$
7. $7x/3 - 1/5 = 11$
8.* $3x/5 + 1/4 = 11/20$
9. $6 - (1^3/_5)x = 2x - 9$
10.* $3\ ^7/_{11} + (2\ ^{11}/_{12})x = (1\ ^7/_{11})x - 4$
11. $0.34x - 1.65 = 11.71$
12. $0.31x + 0.05 = 17.1 - 0.003x$
13. $^7/_x + {}^1/_3 = {}^1/_{2x}$
14.* $^7/_x = {}^3/_4 - {}^4/_x$
15. $(x - 5)/(x + 3) = {}^1/_9$
16. $(x + 3)/(x + 10) = {}^{32}/_{60}$
17. $11/(x - 4) = 22/(x - 1)$
18.* $2(x - 1)/4 = {}^1/_5$
19. $^1/_{2(x+3)} = {}^7/_6$
20. $4(x + 1)/(x - 1) = {}^7/_3$
21. $x - 2 = {}^3/_x$
22.* $3(x + 1)/2 = {}^3/_x$
23. $(x - 1)/4 = {}^3/_{2x}$
24.* $(x - 3)/5 = 0.5x/(x + 0.5)$

25. What is the concentration of a solution of NADH if the molar absorption coefficient $= 6.3 \times 10^3$ l mol^{-1}cm^{-1} and the absorbance $= 0.155$ in a cuvette with a light path length of 4 cm?
26. For many enzyme-catalysed reactions, where $[S]$ represents the concentration of substrate (reactant), the rate of reaction at a particular concentration of substrate is given by:

$$v = \frac{V_{max}[S]}{K_m + [S]}$$

V_{max} is the maximum rate of reaction and K_m is a constant.

(i) Calculate v when $[S] = 0.2$ mmol/l, $V_{max} = 0.3$ µmol/s and $K_m =$ 400 µmol/l.
(ii) If the $K_m = 0.4$ mmol/l, find the ratio v/V_{max} when $[S] =$ 2.0 mmol/l.
(iii) Find K_m if $v = V_{max}/5$ and $[S] = 0.2$ mmol/l.
(iv) Find $[S]$ if $v = 0.91 \times V_{max}$ and $K_m = 0.4$ mmol/l.
(v) Complete Table 5.1 showing the relationship between $[S]/K_m$ and v/V_{max}.

27. The dissociation of a weak acid can be written as HA <--> $H^+ + A^-$ where HA represents the weak acid, H^+ is a hydrogen ion, and A^- is the 'conjugate base', i.e. the original molecule minus the hydrogen ion. For example, for acetic acid,

$$H_3CCOOH \text{ <--> } H^+ + H_3CCOO^-$$

Dissociation is only slight, so at a particular temperature an equilibrium will be reached. The equilibrium constant for the dissociation (K_a), called the acid dissociation constant or ionisation constant is

Table 5.1 Data for Question 26(v), Chapter 5.

$[S]/K_m$	v/V_{max}
0.01	
	0.091
0.50	
1.0	
	0.75
	0.83
	0.91
100	

$$K_a = \frac{[H^+][H_3CCOO^-]}{[H_3CCOOH]} = 1.8 \times 10^{-5} \text{ mol/1 at } 25°C$$

Calculate the $[H^+]$ of a solution made by using 0.5 mol of acetic acid in 1 litre of water at 25°C.
HINT: If $x = [H^+]$, then it also $= [H_3CCOO^-]$, while $[H_3CCOOH] =$ the original acetic acid concentration minus x.

6 LOGARITHMS

6.1 LOGARITHMS AND EXPONENTS

Exponents were introduced in Chapter 4. In section 4.2 you will have seen that numbers can be written as a combination of a *base* and a raised numeral, the *exponent*. For example, $16 = 2^4$ where the base is 2 and the exponent is 4.

The logarithm (log) of a number has the same value as the exponent: the power to which a base must be raised so that it equals the given number.

Example 6.1 $\log_2 16 = 4$ because $2^4 = 16$
Read this as 'the log to the base 2 of 16 equals 4 because 2 to the power 4 equals 16'.
Likewise, $\log_{10} 100 = 2$ because $10^2 = 100$

In general, then, $\log_b a = x$ when $b^x = a$

Since the value of a logarithm depends on the base, this must be **specified** or **implied**. 'Logarithm' is usually abbreviated to 'log' with the specified base as a subscript, e.g. $\log_2$. However, the most frequently encountered logs are those with base 10 (called common logs) and those with base e (called natural or Naperian logs: e being a number with a value of approximately 2.718. It is not necessary to understand how it is derived to use it.) A common log is usually written as 'log', and a natural log as 'ln'. This is how they appear on calculator keys. So,

$\log_2 a$ specifies the base 2
whereas $\log a$ implies $\log_{10} a$ (a common log)
and $\ln a$ implies $\log_e a$ (a natural log)

Note that any positive numbers can have logs, but zero and negative numbers cannot. What is $\log_{10} 0$? It means $10^x = 0$ but there is no value of x for which this could be true: a positive or negative value of x gives a value greater than 0, and $10^0 = 1$. What is $\log_{10}(-10)$? $10^x = -10$. Again, there is no value of x for which this could be true.

You will remember from section 4.2 that when multiplying numbers expressed in exponential form, you add the exponents. Therefore the log of two exponential numbers (having the same base) multiplied together equals the sum of the logs of the numbers.

Example 6.2 $10^2 \times 10^3 = 10^{2+3} = 10^5$
So $\log(10^2 \times 10^3) = \log(10^5)$
$= \log(10^{2+3}) = \log 10^2 + \log 10^3$

The general rule is:

$$\mathbf{\log(a \times b) = \log a + \log b}$$

Similarly, when exponential numbers are divided, the exponents are subtracted. Therefore the log of one exponential number divided by another (having the same base) equals the difference between the logs of the numbers.

Example 6.3 $10^5/10^3 = 10^2 = 10^{5-3}$
So $\log(10^5/10^3) = \log(10^2)$
$= \log(10^{5-3}) = \log 10^5 - \log 10^3$

The general rule is:

$$\mathbf{\log(a/b) = \log a - \log b}$$

At the beginning of this chapter you saw that
$\log_b a = x$, when $a = b^x$
But if $a = b^x$, then $(a)^n = (b^x)^n$ is also true
And $\log_b(a)^n = \log_b(b^x)^n$
$= \log_b(b)^{nx} = nx$
if $x = \log_b a$ and $\log_b(a)^n = nx$
then $\log_b(a)^n = n\log_b a$

Thus the general rule is:

$$\mathbf{\log a^n = n \log a}$$

Example 6.4 $\log(100)^3 = ?$
$\log(100)^3 = \log(100 \times 100 \times 100)$
$= \log 100 + \log 100 + \log 100 = 3\log 100$

So far I have used only examples where the logs are whole numbers, but there is no need for this restriction.

Example 6.5 $\log 25 = 1.3979$ because $25 = 10^{1.3979}$

In the example above, 1.3079 is the log of 25, but also 25 is the *antilog* of 1.3979. The antilog of a number is the value obtained when the base is raised to a power equal to the number to be antilogged.

Example 6.6 $\log 10^5 = 5$, and antilog $5 = 10^5$

Note that this means **the antilog of the log of a number is the number itself.**

If $a = b^x$, $\log_b a = x$
but $\text{antilog}_b x = b^x$
so $\text{antilog}_b(\log_b a) = b^x = a$
or, $\text{antilog}(\log a) = a$

Two other rules follow:

(1) $\mathbf{\log {}^1/_a = -\log a}$
If $x = \log a$, then $a = 10^x$
So ${}^1/_a = {}^1/_{10^x} = 10^{-x}$
and $\log {}^1/_a = \log(10^{-x}) = -x$
$= -\log a$

(2) $\mathbf{(\log \sqrt[n]{a} = {}^{(\log a)}/_n}$
$\sqrt[n]{a}$ means the nth root of a, i.e. the number which when multiplied by itself n times equals a, so it is an inverse power of a.
$\sqrt[n]{a} = a^{1/n}$
and $\log \sqrt[n]{a} = (\log a)^{1/n} = {}^{(\log a)}/_n$

Example 6.7 $\log \sqrt[3]{8} = {}^{(\log 8)}/_3$
As $8 = 2 \times 2 \times 2 = 2^3$
$\sqrt[3]{8} = 2 = 8^{1/3}$
and $\log \sqrt[3]{8} = \log 2 = \log 8^{1/3} = {}^{(\log 8)}/_3$

6.2 MANIPULATING LOGS

Using the rules for logs explained above, it is possible to rearrange, simplify, or solve equations containing logs. The rules are restated here in various useful forms. For any positive non-zero number a and any positive non-zero number b,

(1) $\mathbf{\log(ab) = \log a + \log b}$
(2) $\mathbf{a \times b = \text{antilog}(\log a + \log b)}$
(3) $\mathbf{\log(a/b) = \log a - \log b}$
(4) $\mathbf{a/b = \text{antilog}(\log a - \log b)}$
(5) $\mathbf{\log {}^1/_a = -\log a}$
(6) $\mathbf{\log a^n = n \log a}$
(7) $\mathbf{\log \sqrt[n]{a} = {}^{(\log a)}/_n}$

Example 6.8 Solve $\log_2(x-4) = 3$
Taking antilogs of both sides
$\text{antilog}(\log_2(x-4)) = \text{antilog}_2 3$
So $x - 4 = 2^3 = 8$
$x = 12$

Example 6.9 Express $y = ax^{-c}$ in natural logarithmic form
Following rule (1)
$\ln y = \ln a + \ln x^{-c}$
Following rule (6)
$\ln y = \ln a - c \ln x$

Example 6.10 Using $\log 2 = 0.3$ and $\log 3 = 0.48$, calculate the following, without using a calculator.

(i) $\log 2^3$
Using rule (6)
$\log 2^3 = 3\log 2$
$= 3(0.3) = 0.9.$
(ii) $\log {}^2/_3$
Using rule (3)
$\log {}^2/_3 = \log 2 - \log 3$
$= 0.3 - 0.48 = -0.18$
(iii) $\log 6$
Using rule (1)
$\log 6 = \log(2 \times 3)$
$= \log 2 + \log 3$
$= 0.3 + 0.48 = 0.78$

Example 6.11 Express $\ln x$ in terms of $\log x$.
Remember that the natural log, ln, has the base e.
If $\ln x = a$, $x = e^a$
Using rule (6)
$\log x = \log(e)^a = a\log e$
so $a = \log x/\log e = \log x/0.434 = 2.303\log x$
$\ln x = 2.303\log x$

Example 6.12 Find n when $9 = 8^n$
Using rule (6)
$\log 8^n = n\log 8$
so $n = \log 9/\log 8 = 1.057$

You can check this is correct using your calculator if you have a key labelled 'x^y'. Press: 8, x^y, 1.057, =. You may have to use a supplementary key to obtain x^y, e.g. the INV key, the second function key, or the shift key in which case the sequence would be, as an example, 8, INV, x^y, 1.057, =.

Example 6.13 Find n when $n = (1.8)^{2.5}/(2.7)^{3.2}$

Following rule (3),

$$\log n = \log\left(\frac{(1.8)^{2.5}}{(2.7)^{3.2}}\right) = \log(1.8)^{2.5} - \log(2.7)^{3.2}$$

$$\log n = 2.5 \log 1.8 - 3.2 \log 2.7$$
$$= 2.5(0.255) - 3.2(0.431)$$
$$= -0.7417$$
so $n = \text{antilog}(-0.7417) = 0.181$

Example 6.14 Solve $\log_2(x - 4) = 3$

$\text{antilog}(\log_2(x - 4)) = \text{antilog}_2 3$

$x - 4 = 2^3 = 8$

$x = 12$

Example 6.15 Find $\sqrt[3]{20}$

Following rule (7)

$$\log \sqrt[3]{20} = \frac{\log 20}{3}$$

$$\text{so } \sqrt[3]{20} = \text{antilog}\left(\frac{\log 20}{3}\right)$$
$$= \text{antilog}(0.4337) = 2.714$$

Note that if your calculator has an '$x^{1/y}$' button, you could calculate this directly. Press: 20, INV (if necessary), $x^{1/y}$, 3, =.

In Chapter 5, problem 27 involved the calculation of the $[H^+]$ in a solution of a weak acid with an acid dissociation constant represented by K_a. Because of the sizes of the values for these it is convenient to transform these numbers by taking logs, and inventing a term for these transformed numbers. Thus, the 'pH' of a solution is defined as the negative logarithm to the base 10 of its hydrogen ion concentration, i.e.

$$\mathbf{pH = -log[H^+]}$$

Similarly, the pK_a of a weak acid is defined as the negative logarithm to the base 10 of its acid dissociation constant, i.e.

$$\mathbf{pK_a = -log\ K_a}$$

Water itself dissociates, so it also has a K_a value which at 25°C is 1.8×10^{-16} mol/l.

$$H_2O \text{ <--> } H^+ + OH^-$$

More correctly, $$H_2O + H_2O \text{ <--> } H_3O^+ + OH^-$$

but for most purposes you can think of a hydrogen ion being formed rather than a hydronium ion (H_3O^+), so

$$K_a = \frac{[H^+][OH^-]}{[H_2O]} = 1.8 \times 10^{-16} \text{ mol/l}$$

Pure water is 55.6 mol/l, ignoring the negligible reduction in concentration caused by ionisation, so

$$[H^+][OH^-] = K_a[H_2O] = 1.8 \times 10^{-16} \times 55.6 = 10^{-14} \text{ (mol/l)}^2$$

This is known as K_w, the ion product of water.

Since $[H^+] = [OH^-]$
$[H^+] = \sqrt{10^{-14}} = 10^{-7}$ mol/l
Expressing these value in terms of logs,
$[H^+][OH^-] = 10^{-14}$
$\log([H^+][OH^-]) = \log 10^{-14} = -14$
$\log[H^+] + \log[OH^-] = -14$
$-pH + \log[OH^-] = -14$
$\log[OH^-] = -14 + pH$
$-\log[OH^-] = 14 - pH$
But $-\log[OH^-]$ is defined as pOH
so pOH = 14 – pH
and pH = pOH – 14

Using this relationship it is possible to calculate the pH of a solution of a base.

Example 6.16 What is the pH of a 0.2 mol/l aqueous solution of NaOH at 25° C ?

NaOH is a strong base which completely dissociates. It will give rise to 0.2 mol/l OH^-, so $[OH^-] = 0.2$ mol/l.

$pOH = -\log[OH^-] = -\log(0.2) = 0.699$
So $pH = 14 - pOH = 14 - 0.699 = 13.3$.

The Henderson-Hasselbalch equation expresses the relationship between pH, pK_a, and the ratio of the concentrations of conjugate base and weak acid when a weak acid is neutralised by a strong base.

$$\mathbf{pH = pK_a + \log\frac{[conjugate\ base]}{[acid]}}$$

Example 6.17 Check that the answers obtained in problem 27 in Chapter 5 satisfy this equation.

$[H^+] = 2.99 \times 10^{-3}$ mol/l

$pH = -\log[H^+] = -\log(2.99 \times 10^{-3}) = 2.52$

$K_a = 1.8 \times 10^{-5}$ mol/l

$pK_a = -\log(1.8 \times 10^{-5}) = 4.74$

$[\text{conjugate base}] = [H^+] = 2.99 \times 10^{-3}$ mol/l

$[\text{acid}] = 0.5$ mol/l $- 2.99 \times 10^{-3}$ mol/l $= 0.497$ mol/l

Does $pH = pK_a + \log\frac{[\text{conjugate base}]}{[\text{acid}]}$?

$$2.52 = 4.74 + \log\frac{(2.99 \times 10^{-3})}{(0.497)}$$

$$= 4.74 + \log(6.02 \times 10^{-3})$$

$$= 4.74 - 2.22$$

$$= 2.52$$

The Henderson-Hasselbalch equation is particularly useful in calculating the amounts of acid and conjugate base required to make a buffer solution of a certain pH.

Example 6.18 How could you prepare 1 litre of 0.1 mol/l sodium phosphate buffer pH 7.2, given solutions of 0.1 mol/l Na_2HPO_4 and 0.1 mol/l NaH_2PO_4? The pK_a value for the dissociation $H_2PO_4^- <--> HPO_4^{2}- + H^+$ is 6.8.

$$pH = pK_a + \log\frac{[\text{conjugate base}]}{[\text{acid}]}$$

$$7.2 = 6.8 + \log\frac{[HPO_4^{\ 2-}]}{[H_2PO_4^{\ -}]}$$

$$0.4 = \log\frac{[HPO_4^{\ 2-}]}{[H_2PO_4^{\ -}]}$$

$$\frac{[HPO_4^{\ 2-}]}{[H_2PO_4^{\ -}]} = \text{antilog } 0.4 = 2.51$$

The proportion is therefore 2.51 volumes of 0.1 mol/l Na_2HPO_4 to 1 volume of 0.1 mol/l NaH_2PO_4.

So you need $(^{2.51}/_{3.51}) \times 1$ litre = 0.715 litres of 0.1 mol/l Na_2HPO_4 plus 0.285 litres of 0.1 mol/l NaH_2PO_4. Add them and mix.

The equation can also be used to calculate the proportion of molecules of a weak acid or base that are ionised at a certain pH.

Example 6.19 Glycine has the structure $H_3N\text{-}CH_2\text{-}COOH$. The p$K_a$ for the carboxyl group is 2.4. What proportion of the glycine molecules in a dilute aqueous solution have ionised carboxyl groups at pH 3.4?

$$\text{pH} = \text{p}K_a + \log\frac{[\text{conjugate base}]}{[\text{acid}]}$$

$$\frac{[\text{conjugate base}]}{[\text{acid}]} = \text{antilog}(\text{pH} - \text{p}K_a)$$

$$= \text{antilog}(3.4 - 2.4)$$

$$= \text{antilog}(1.0)$$

$$= 10$$

Therefore at pH 3.4, there will be 10 molecules of glycine with an ionised carboxyl group for every one with the group un-ionised.

Therefore the proportion ionised = 10 ionised/11 total = 0.91 or 91%.

6.3 PROBLEMS

If you have just finished reading this chapter I suggest you now try the 'pure mathematics' examples marked '*'. If you get these correct, try *all* of the applied examples. If you also get these correct, move on to the next chapter. If you do not get the right answers, re–read the explanations and try the rest of the problems 1–54, and then repeat your attempts at the applied problems.

Find the values of the following without using a calculator.

1. $16^{1/2}$
2. * $27^{2/3}$
3. $\log 100$
4. * $\log 10^4$
5. $\ln 1$
6. $\ln e^5$
7. $\log 0.0001$
8. * $\log(^1/_{1000})$
9. $\log 10^{-7}$
10. $\ln (e^2)^3$
11. $\log_2 16$
12. * $\log_5 125$
13. $\log_7(^1/_{49})$
14. $\log_8 32$
15. $\log_{25} 5$
16. * $\log_{32} 8$ (Hint: think in terms of base 2)

Using log 2 = 0.30 and log 3 = 0.48, and without using a calculator, find the values of the following.

17. $\log 4$
18. $\log 36$
19. $\log 0.6$
20.* $\log 1.5$
21. $\log (2/3)$
22. $\log 5$
23. $\log 200$
24.* $\log(3 \times 10^4)$
25. $\log (6 \times 10^{-3})$
26.* $\log 0.04$

Solve the following equations.

27. $2^{x+1} = 8$
28. $10^{3x-1} = 5$
29. $9^{x+3} = 3^x$
30.* $10^{(x^2)} = 16$
31. $\log_4 x = 2$
32. $\log_3 x = 4$
33. $\log_2 x = 7$
34.* $\log_5 x = -3$
35. $\log_4 x = -2$
36. $\log_4 x = 1/2$
37. $\log_{16} x = 3/4$
38. $\log_{10} x = 3^2$
39. $\log_{10} x = 10^{-3}$
40.* $\log_{10} x = -1/100$
41. $\log_x 8 = 3$
42. $\log_x 169 = 2$
43. $\log_x 64 = -2$
44.* $\log_x(1/125) = -3$
45. $\log x + \log 5 = -1$
46. $\log x - \log 0.01 = 0.1$
47. $\log(x + 1) + \log 5 = 4$
48.* $\log(x - 2) - \log 4 = 3$
49. $\log(x + 1) + \log(x - 1) = 3$
50. $\log(x + 1) - \log(x - 3) = 7$
51.* $\log(x + 1) + \log(x - 4) = \log 6$
52. $\log(2x + 1) - \log x = \log(x + 2)$

53.* Express in natural logarithmic form the equation $y = ax^b$

54. Express in exponential form the equation $\log y = a\log x + b$.

55. You can set a spectrophotometer to read either transmittance (T) or absorbance (A). Percentage transmittance is the percentage of the incident light that passes through the sample. Absorbance is related to transmittance by the equation:

$$A = -\log T.$$

(i) Complete Table 6.1.
(ii) What is the transmittance of a solution with an absorbance of 0.5?

56. Calculate the pH values of solutions with the following H^+ concentrations.
(i) 0.1 mol/l
(ii) 0.0001 mol/l
(iii) 5×10^{-3} mol/l
(iv) 8×10^{-11} mol/l
(v) 6×10^{-3} mmol/l

57. Calculate the pH values of solutions with the following OH^- concentrations.
(i) 0.1 mol/l
(ii) 0.001 mol/l

Table 6.1 Values for Question 55, Chapter 6.

Percentage T	*A*
100	
	1
	2
0.1	

(iii) 3×10^{-3} mol/l
(iv) 1.7×10^{-5} mmol/l
(v) 11×10^{-8} mol/l

58. Calculate the H^+ and OH^- concentrations of solutions with the following pH values.
    (i) 2.1
    (ii) 8.0
    (iii) 3.6
    (iv) 11.4
59. Ammonia is a weak base that in aqueous solution forms its conjugate acid $NH_4{}^+$ by reaction with water:

$$NH_3 + H_2O <--> NH_4{}^+ + OH^-$$

The pK_a for the $NH_4{}^+$ is 9.25. What will be the pH of a 0.01 mol/l solution of NH_3? (Hint: think of the reverse reaction.)
60. The K_a of the conjugate acid of the base ethylamine is 1.58×10^{-11} mol/l. Which base, ammonia or ethylamine, is stronger, i.e. produces a solution with the higher pH for the same concentration?
61. What is the pH of a solution made by adding 100 ml of 0.05 mol/l acetic acid to 200 ml of 0.01 mol/l sodium acetate? (pK_a of acetic acid is 4.74.)
62. If the amino acid alanine has pK_a values of 2.3 for the carboxyl group and 9.7 for the amino group, what proportion of the carboxyl and amino groups will be ionised at pH 7.4?
63. If the pH of blood is 7.4 and the total concentration of inorganic phosphate (consisting of $PO_4{}^{3-}$, $HPO_4{}^{2-}$, $H_2PO_4{}^-$ and H_3PO_4) is 1 mmol/l, what are the approximate concentrations of the ions $PO_4{}^{3-}$, $HPO4^{2-}$,and $H_2PO_4{}^-$. (Use pK_a values of 2.14, 6.86 and 12.4.)
64. The equilibrium constant for a reaction is related to the standard free energy change ΔG^o by the equation:

$$K_{eq} = e^{-\Delta G^\circ/RT}$$

where R = 8.3 J K^{-1} mol^{-1} and T is temperature (kelvin). Calculate ΔG^o for a reaction when the temperature is 37° C and K_{eq} = 0.50.

7 STRAIGHT-LINE GRAPHS: CALIBRATION CURVES AND LINEAR RATES OF CHANGE

7.1 PROPORTIONAL RELATIONSHIPS

In previous chapters you have come across many examples where one number is related to another in a very simple way. For example, the circumference of a circle is π times the diameter and the absorbance of a solution is a constant times the concentration. These relationships can be expressed as:

$c = \pi d$, where c is the circumference, and d is the diameter, and
$A = \mathrm{k}c$, where A is the absorbance, k is a constant and c is the concentration.

Note that each of these can be rearranged as a proportion, i.e. so that the value of one variable is proportional to the value of the other:

$$c/d = \pi \quad \text{and} \quad A/c = \mathrm{k}$$

In experiments we are often trying to find out the relationship between two variables, i.e. to answer questions of the type. 'How does the value of one variable, y, change when the value of the other variable, x, is changed'. An experiment to answer this question is likely to produce a set of data consisting of values of x, each of which has a corresponding value of y. One way of analysing this data to show the relationship between x and y is to plot them on a graph in which the horizontal distance represents the value of x, and the vertical distance the value of y. Since each experimental value of x has an associated value of y, it can be represented by a point whose coordinates are those values. Any point, with the coordinates x_n and y_n, can be written as $P_n(x_n, y_n)$.

Example 7.1 The data point P_1 where $x = 2$ and $y = 3$ is plotted as shown in Graph 7.1. Note that there are x and y axes having scales starting at 0 and that each unit of distance along the axis represents the same increase in value. Remember that the horizontal x coordinate is always given first, so the point can be labelled $P_1(2, 3)$.

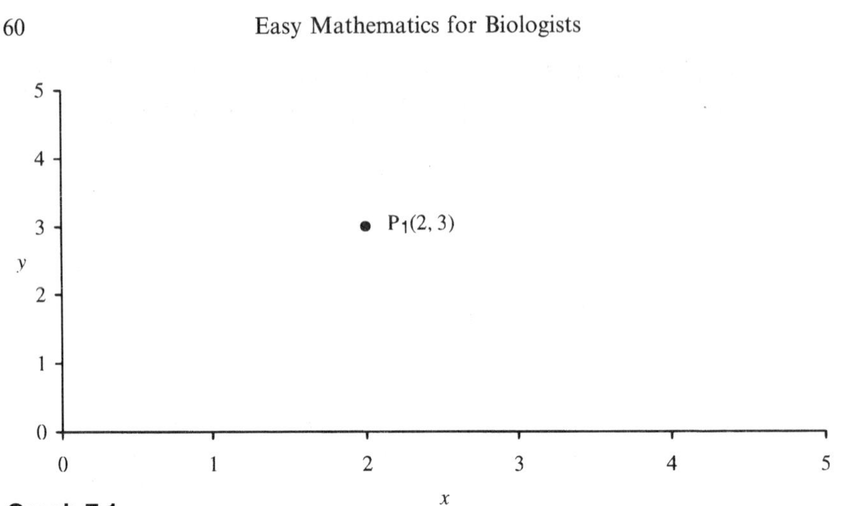

Graph 7.1

Example 7.2 From an experiment the results shown in Table 7.1 were obtained. The data can be plotted on a graph as shown in Graph 7.2.

Each point on this graph represents a pair of values of x and y. Note that in this example, all the points lie on a straight line which passes through the origin (the point (0, 0)). Taking the point (0.100, 0.6) you can see that the value of y is 6 times the value of x ($0.6 = 6 \times 0.100$). If you look at any other point, you will see that this is still true. Check for yourself using the original data points and then some intermediate points. Since the value of y is always 6 times the value of x, the relationship can be written as:

$$y = 6x$$

which is an equation of the form $\boldsymbol{y} = \mathbf{a}\boldsymbol{x}$ where a is a constant, and the value of y is proportional to the value of x, i.e. $y/x = \mathrm{a}$.

On this same graph, y/x is the steepness of the straight line, the ratio of the vertical *change* to the horizontal *change*. This is usually known as

Table 7.1 Data for Example 7.2.

x	y
0.0	0.0
0.025	0.15
0.050	0.30
0.075	0.45
0.100	0.60

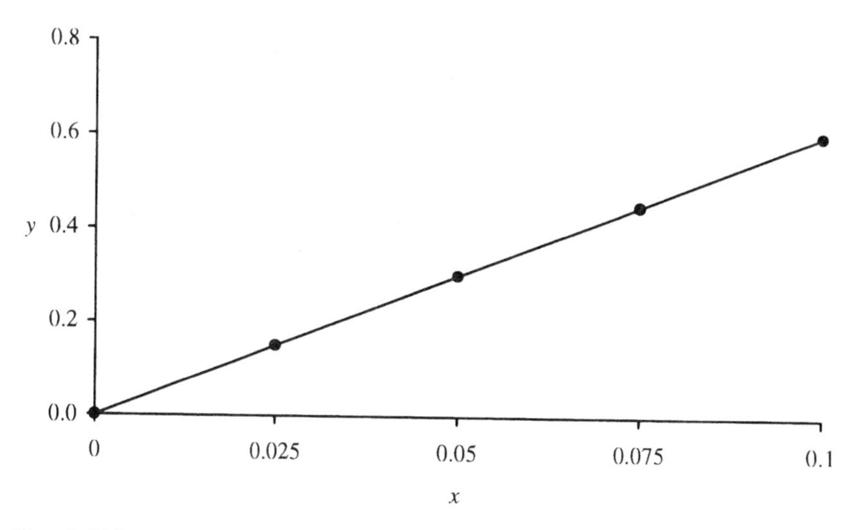

Graph 7.2

the **gradient** or **slope** of the line. If you start at $P_1(x_1, y_1)$ and move along the line to $P_2(x_2, y_2)$, you will have moved vertically by $y_2 - y_1$ and horizontally by $x_2 - x_1$, so the

$$\textbf{gradient} = (y_2 - y_1)/(x_2 - x_1).$$

(This is also sometimes written as $\Delta y/\Delta x$ where delta y and delta x denote the change in $\boldsymbol{y}$ and $\boldsymbol{x}$. The gradient is therefore a measure of how many units y changes for a change of 1 unit in x.

Example 7.3 Find the value of the gradient in Graph 7.3.
You could obtain the gradient by choosing two points e.g. (0.100, 0.60) and (0.050, 0.30) and calculating the gradient as:

$$\text{gradient} = \frac{0.60 - 0.30}{0.100 - 0.050} = \frac{0.30}{0.050} = 0.60$$

Note that it does not matter which point is P1 and which is P2:

$$\text{gradient} = \frac{0.30 - 0.60}{0.050 - 0.100} = \frac{-0.30}{-0.050} = 0.60$$

Example 7.4 Find the gradient of the straight line passing through the points $P_1(1, 3)$ and $P_2(3, 9)$.

$$\text{gradient} = \frac{9 - 3}{3 - 1} = \frac{6}{2} = 3$$

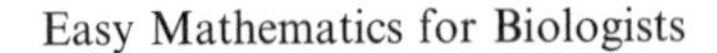

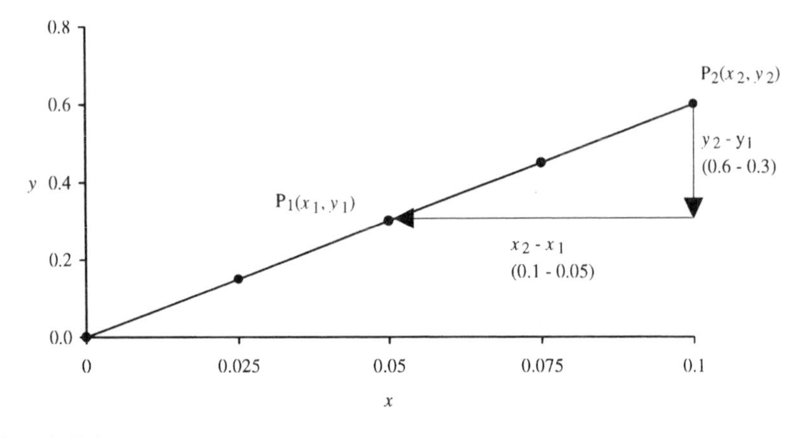

Graph 7.3

7.2 EQUATIONS OF STRAIGHT LINES

Look at Graph 7.4. How does y change for a change in x? Choose any two points e.g. (0, 2) and (5, 5). The ratio $^{\text{change in } y}/_{\text{change in } x} = (5 - 2)/(5 - 0) = {}^{3}/_{5} = 0.6$. As this is true for any two points, the **change** in $\boldsymbol{y}$ is proportional to the **change** in $\boldsymbol{x}$, but in this case $\boldsymbol{y}$ **is not proportional to** $\boldsymbol{x}$. This is because the line does not pass through the origin, the point where both x and y are zero. The line cannot therefore be represented by the equation $y = 0.6x$, because, for example, if you substitute the coordinates of

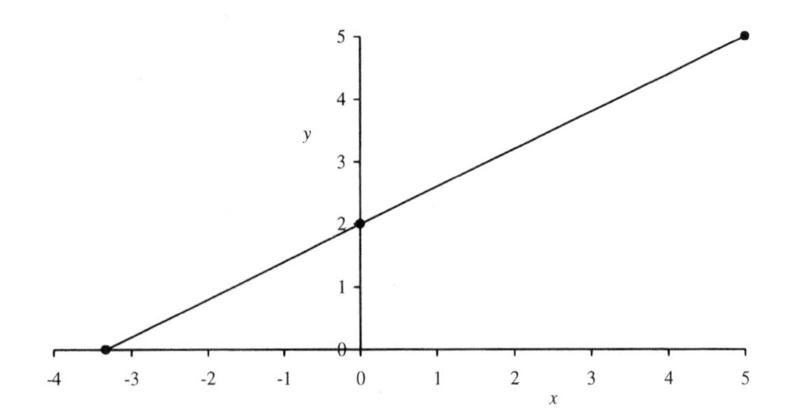

Graph 7.4

the point (5, 5) in the equation you get $5 = 0.6 \times 5 = 3$ which is nonsense. If you look again at the graph you can see that when $x = 0, y = 2$, so the correct equation is $y = 0.6x + 2$.

In general, any straight line can be represented by an equation which can be expressed in the form $\boldsymbol{y = ax + b}$ where a and b are constants. a is the gradient and b, the value of y when $x = 0$, is called the ***y*-intercept**. Both $\boldsymbol{a}$ and $\boldsymbol{b}$ can have positive, zero, or negative values.

Example 7.5 $-3x + 6y = 9$

Simplifying gives

$-x + 2y = 3$

$2y = x + 3$

$y = 0.5x + 1.5$

In this case, the gradient is 0.5 and the y-intercept is 1.5.

In Graph 7.5, the y-intercept $b = -1$ as the line intercepts the y-axis at the point $(0, -1)$. The gradient is $(0 - (-1))/(2 - 0) = {}^1/_2 = 0.5$. Therefore the equation of this line must be $y = 0.5x - 1$.

In Graph 7.6, the y-intercept is the point (0, 5) and the slope is $(0 - 5)/(5 - 0) = -5/5 = -1$, so the equation of this line must be $y = -x + 5$.

In Graph 7.7 the line is parallel to the x-axis so the value of y does not change whatever the value of x, so the gradient = 0. The value of y is said to be *independent* of x. Since the y-intercept is the point (0, 2) the equation of the line must be $y = 2$.

In Graph 7.8, the line is parallel to the y-axis and there is no y-intercept. The line represent a set of values of y which have no defined relationship to x, so the gradient is undefined and the equation of this line is $x = -2$.

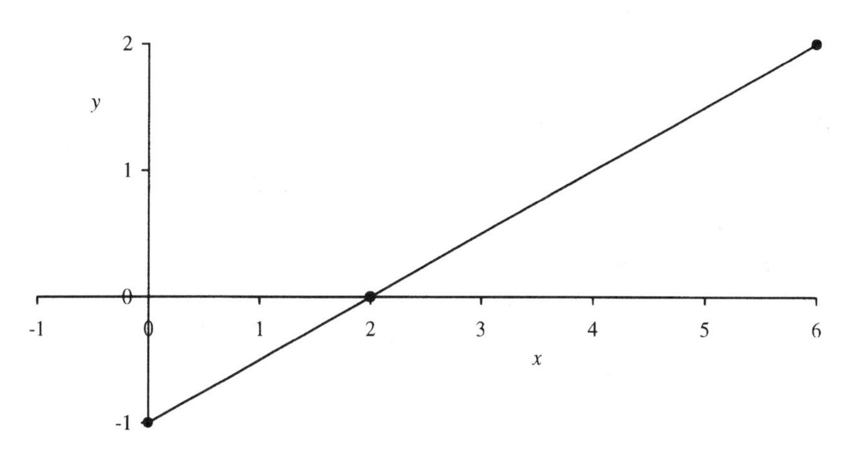

Graph 7.5

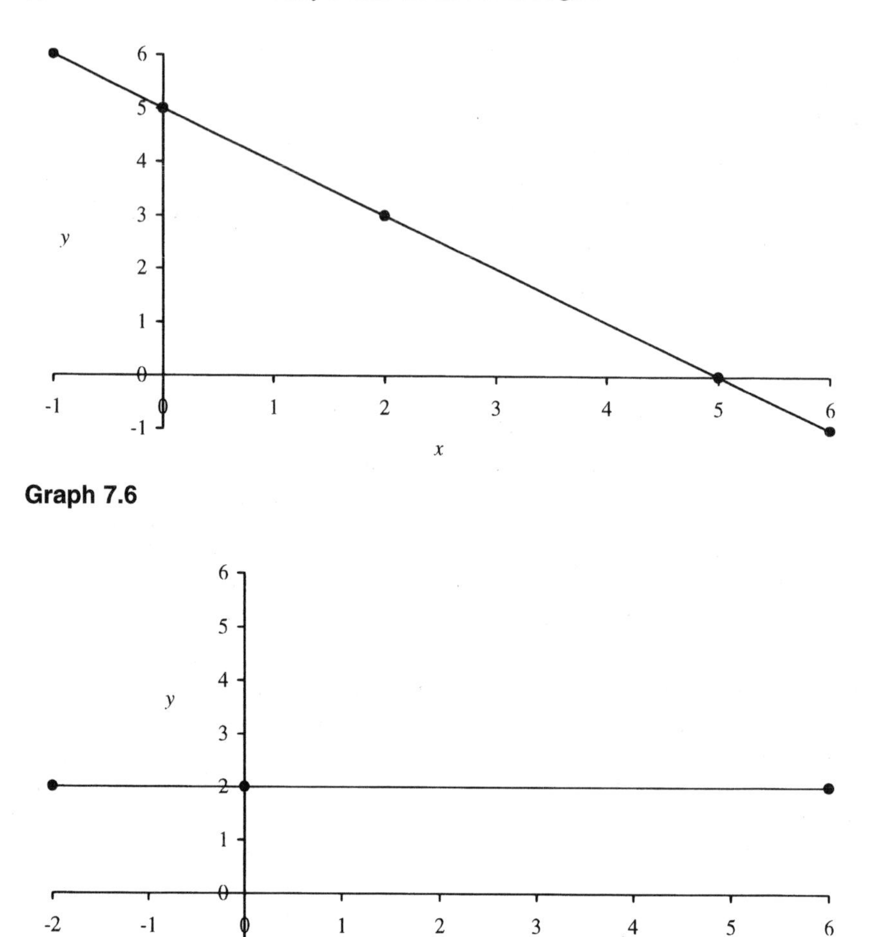

Graph 7.6

Graph 7.7

It is possible to determine the equation of a straight line given either

- the coordinates of two points on it, or
- the coordinates of one point plus the gradient, or
- the gradient and the y-intercept.

Example 7.6 Determine the equation of the straight line passing through the points $(-3, 7)$ and $(5, 11)$.

Gradient $= (11 - 7)/(5 - (-3)) = {}^4/_8 = 0.5$

Substituting for a in $y = ax + b$

$y = 0.5x + b$

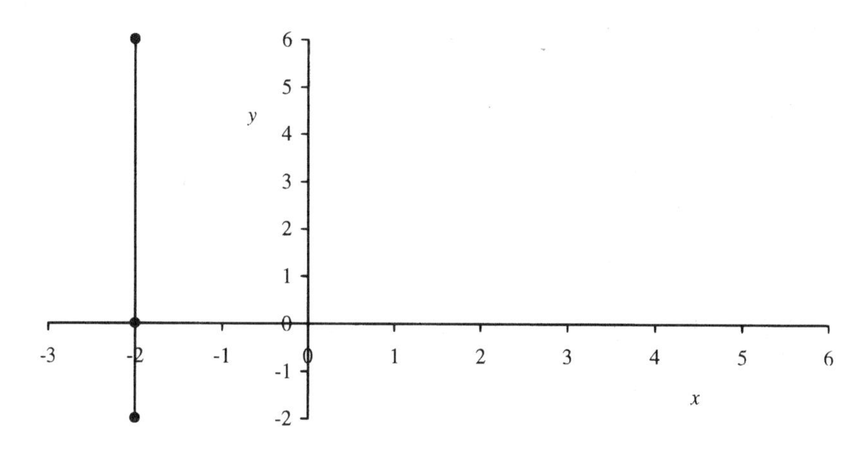

Graph 7.8

Using the values of y and x at one point
$11 = 0.5(5) + b$
$11 = 2.5 + b$
$b = 8.5$
Therefore $y = 0.5x + 8.5$.

Example 7.7 Determine the equation of the straight line passing through the point $(3, 4)$ with a gradient of -5.
The equation will be of the form $y = ax + b$, where $a = -5$ substituting the values of y and x
$4 = -5(3) + b$
$4 = -15 + b$
$b = 19$
so $y = -5x + 19$

Example 7.8 Determine the equation of the straight line with a gradient of -3 and y-intercept of -4.
$y = ax + b$ where a is the gradient and b the y-intercept,
so $y = -3x - 4$

Given the equation of a straight line you can determine the coordinates of two points by substituting values for x.

Example 7.9 $y = -5x + 20$
If $x = 1$, then $y = -5 + 20 = 15$ so one point is (1, 15)
If $x = 2$, then $y = -10 + 20 = 10$ so another point is (2, 10).

Using these coordinates you could draw a graph with the straight line passing through these points.

An alternative is to find the x and y intercepts.

When $x = 0$, $y = 20$ so the y-intercept is (0, 20)

When $y = 0, 5x = 20, x = 20/5 = 4$ so the x-intercept is (4, 0).

7.3 EQUATIONS OF STRAIGHT LINES AND STRAIGHT-LINE GRAPHS IN PRACTICE

If you carry out an experiment in which you change the value of x and determine the corresponding value of y, you can plot the values of x and y on a graph. You should make sure you do the following six things:

(a) Choose appropriate scales and mark them on the axes. The scales will normally be linear; for example 1 cm along the axis might represent a change of 1 minute wherever you start from on the axis. You should normally include the origin on your graph. If you do not, you should make it obvious that this is the case, perhaps by having a break in the axis.

(b) Label the axes to indicate what variables are being plotted. The units on the x-axis are those of the independent variable. This is the one whose values may have been chosen, or do not depend on the other variable. The dependent variable (the one whose value is determined by the other variable) is plotted on the y-axis.

(c) Label the axes with appropriate units.

(d) Mark the points clearly.

(e) Decide whether or not the points lie on a straight line. This is not always easy. Because of experimental error or natural variation the data points are unlikely to lie exactly on a straight line. Also, in some cases, there may be a proportional relationship between the variables for only part of the range of values of x. These points are illustrated in Graphs 7.9 and 7.10. Graph 7.9 shows a situation where y is proportional to x only over part of the range of x-values. In Graph 7.10 there is some doubt whether a straight-line relationship really exists. Graphs of curves are dealt with in Chapter 8.

(f) If you are convinced that there is a straight-line relationship between x and y you need to draw the line which best fits the points. You can do this by eye, trying to balance points above and below the line, or you can use a statistical method (linear regression) which is available on many

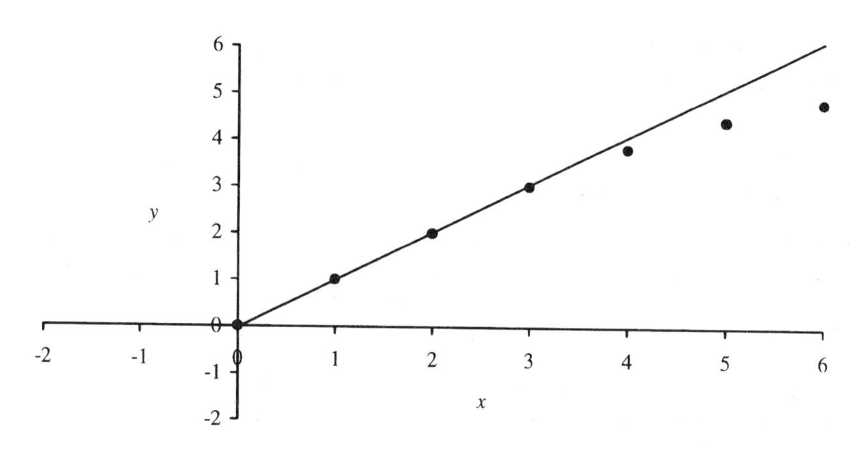

Graph 7.9

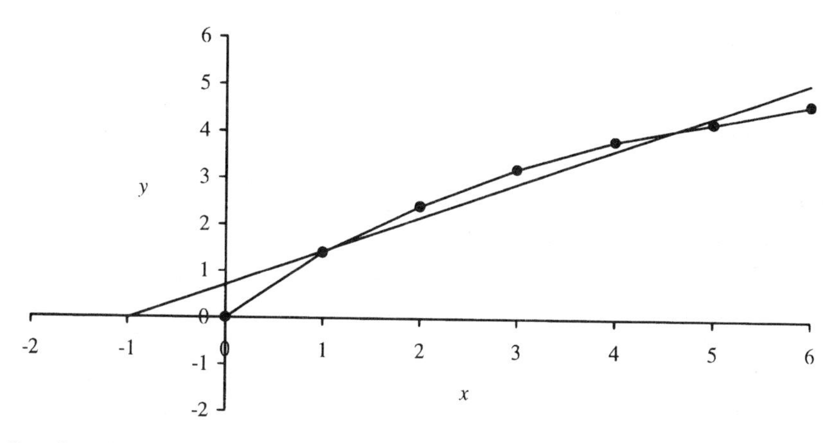

Graph 7.10

calculators and in computer statistical or data presentation packages such as MINITAB. (Statistical methods can be used to determine the goodness-of-fit of the data to a straight-line relationship.) Having drawn the line you can determine the gradient and y-intercept, and hence the equation of the line.

You can interpret a straight-line graph as follows:

(a) The variable y is related to x such that any increase in x is associated with a proportional increase in y. The gradient of the line gives the ratio between the change in y and change in x.

(b) If the line passes through the origin, y is proportional to x.
(c) If the line does not pass through the origin, when $x = 0$, y has a non-zero value.

You may be able to predict values of y for another set of values of x, or use measured values of y to determine x.

Example 7.10 The results shown in Table 7.2 were obtained when the intensity of the light emitted from a set of solutions containing different known concentrations of Ca^{2+} (a set of 'standards') was measured in a flame photometer.

Table 7.2 Data for Example 7.10.

Ca^{2+} concentration (μg/ml)	*Reading*
0	0
20	11
40	20
60	29
80	40
100	50

If the readings for these standards are plotted against their Ca^{2+} concentrations (µg/ml), a straight-line graph is obtained which passes through the origin and has a slope of 0.5 (see Graph 7.11).

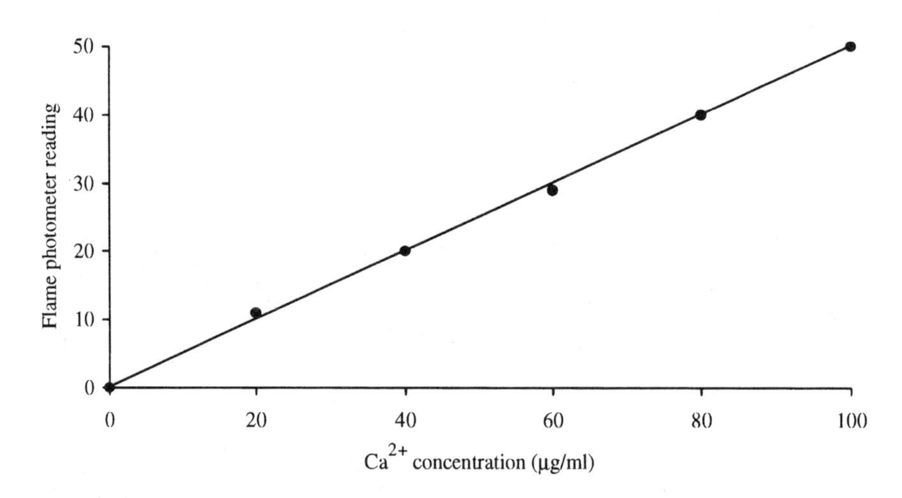

Graph 7.11

The equation is therefore

$$\text{reading} = 0.5(\text{concentration}(\mu g/ml)).$$

This indicates that the reading is proportional to the Ca^{2+} concentration over the range used. If readings are obtained from other solutions (samples with unknown Ca^{2+} concentrations), their concentrations can be calculated by dividing the reading by the gradient.

$$\text{reading} = 0.5(\text{concentration})$$
$$\text{so sample reading}/0.5 = \text{concentration } (\mu g/ml).$$

This example illustrates the use of a 'calibration curve' which is a general term for the graph obtained when the readings from an instrument are plotted against the known values of another variable.

In many experiments time is the independent variable (x), in which case the gradient of the straight-line graph represents the **rate** of change of the other variable. Suppose you are measuring the amount of product obtained from a reaction at various times after starting the reaction. When you plot the amount of product against time, you find the points lie on a straight line passing through the origin. You are therefore able to conclude that product accumulation is proportional to time over the range measured. The gradient gives the **rate of reaction**, i.e. the **rate** at which product is made. If you repeated this experiment at a higher temperature, you might find that the gradient increased. You could conclude that the rate of reaction depends on the temperature.

Another example would be if you were measuring the rates of reaction using different concentrations of enzyme. For each enzyme concentration you would plot a graph of product accumulated against time, and obtain the gradient of the straight line. These gradients represent the rates of reaction at different enzyme concentrations. When you plot these rates of reaction against enzyme concentration you find the points lie on a straight line passing through the origin. You are therefore able to conclude that under the conditions used, the rate of reaction is proportional to the enzyme concentration.

7.4 PROBLEMS

If you have just finished reading this chapter I suggest you now try the 'pure mathematics' examples marked '*'. If you get these correct, try *all* of the applied examples. If you also get these correct, move on to the next chapter. If you do not get the right answers, re-read the explanations and try the rest of the problems 1–16, and then repeat your attempts at the applied problems.

Find the gradient of the straight line passing through the points whose coordinates are given.

1. $(2, 3), (5, 6)$
2.* $(-1, 3), (-5, -6)$
3. $(0, -5), (6, 5)$
4. $(0.25, 4), (4, -1.5)$

Find the equation of the straight line passing through the points whose coordinates are given.

5. $(3, 2), (6, 5)$
6.* $(3, 1), (5, 6)$
7. $(2, 3), (-6, 5)$
8.* $(-1, -1), (2, 8)$

Find the equation of the straight line passing through the point P1 with the given gradient *a*.

9. $P1(-1, 2)$, $a = 2$
10. $P1(-1, -3)$, $a = -1$
11. $P1(-1, 4)$, $a = \frac{1}{4}$
12.* $P1(5, -3)$, $a = -\frac{3}{5}$

Graph the following equations:

13. $y = 2x + 1$
14. $y = -2x - 3$
15. $3 - 2y = 4x$
16.* $4y - 3x = 9$

17. A 0.25 mmol/l stock solution of p-nitrophenol in buffer at pH 10.0 was diluted in buffer to prepare a set of solutions of different concentration. The absorbance at 410 nm of each of these solutions was measured in a spectrophotometer using a 1 cm light path. The results are given in Table 7.3. Calculate the molar absorption coefficient for p-nitrophenol if $A = \mathrm{E}cd$ where A = absorbance, E = molar absorption coefficient, c = concentration, and d = light path length.

Table 7.3 Data for Question 17, Chapter 7.

Volume of stock solution (ml)	*Volume of buffer (ml)*	*Absorbance*
1.0	4.0	0.91
0.8	4.2	0.72
0.6	4.4	0.55
0.4	4.6	0.37
0.2	4.8	0.18
0.0	5.0	0.00

18. Using the data in 17 above, calculate the concentrations of the unknown solutions of p-nitrophenol which when diluted as indicated, gave the absorbances shown in Table 7.4.
19. In an assay for the enzyme lumpase in saliva, the enzyme is incubated with the substrate, L-LUMPA, and the rate of formation of product, BITTA, is measured by taking samples at times during the incubation and adding concentrated HCl which converts BITTA to a coloured substance, MUDDY. The intensity of the colour can be measured spectrophotometrically and is proportional to the concentration of MUDDY. For every molecule of L-LUMPA used, one molecule of MUDDY is produced.

    An experiment was performed as follows:

    1.0 ml saliva + 9.0 ml buffer-L-LUMPA solution were incubated at 37° C. 1.0 ml samples were taken at set times, 2.0 ml conc. HCl added and the absorbance measured against a suitable blank. A 1.0 cm light path was used. The results for this are given in Table 7.5, and those for a calibration curve are given in Table 7.6.

    Calculate:

    (i) The rate of reaction in units of absorbance change minute^{-1}.
    (ii) The rate of reaction in units of μmoles L-LUMPA converted minute^{-1}.
    (iii) The activity of the enzyme in units of μmol minute^{-1} ml^{-1} saliva.

Table 7.4 Data for Question 18, Chapter 7.

Unknown solution	*Volume of unknown (ml)*	*Volume of buffer (ml)*	*Absorbance*
Sample 1	5.0	0.0	0.45
Sample 2	1.0	4.0	0.30
Sample 3	0.5	4.5	0.65
Sample 4	0.3	2.7	0.50

Table 7.5 Data for Question 19, Chapter 7.

Time of incubation (minutes)	*Absorbance*
0	0.00
5	0.10
10	0.20
15	0.30
20	0.40
25	0.50

Table 7.6 Calibration Curve Data for Question 19, Chapter 7.

Tube	*1 mmol/l BITTA (ml)*	*Conc.HCl (ml)*	*Distilled water (ml)*	*Absorbance*
1	0.0	2.0	1.0	0.00
2	0.2	2.0	0.8	0.15
3	0.4	2.0	0.6	0.30
4	0.6	2.0	0.4	0.45
5	0.8	2.0	0.2	0.60
6	1.0	2.0	0.0	0.75

Recently an artificial substrate (JOBBO) for the enzyme has been made which gives a red-coloured product without the need for the addition of conc. HCl. This product has a molar absorption coefficient of 5,000 l $mol^{-1}cm^{-1}$. In an assay with the same saliva sample as before, when 1.0 ml saliva was incubated with 9.0 ml buffer-JOBBO solution at 37°C, the absorbance increased linearly with time from a value of 0 at time = 0 min, to 0.5 at time = 10 min.

(iv) Calculate the activity of the enzyme in the sample.

20. You obtain the results shown in Table 7.7 in an experiment to determine the effect of substrate concentration on the rate of an enzyme-catalysed reaction. Plot a graph of rate of reaction against substrate concentration. What can you say about the relationship between the two variables?

Table 7.7 Data for Question 20, Chapter 7.

Substrate concentration (mmol/l)	*Rate of reaction (μmol/min)*
0.063	0.37
0.084	0.42
0.105	0.46
0.14	0.51
0.25	0.58
0.35	0.63

8 NON-LINEAR RATES OF CHANGE: GRAPHS, TRANSFORMATIONS AND RATES

8.1 GRAPHS THAT ARE NOT STRAIGHT LINES, AND THEIR TRANSFORMATION

Question 20 in the previous chapter had data that when plotted directly gives a curve rather than a straight line. You will often find this to be the case, because the relationship between many variables is not one of direct proportionality.

If the rate of a chemical reaction is proportional to the concentration of the reactant A, for the reaction A → B, the rate of reaction = k[A] where k is a constant (called the rate constant). Also, the rate of reaction = the rate of disappearance of A which can be represented by the expression $-{}^{d[A]}/_{dt}$ where the 'd' stands for 'a small change in', and the minus sign is present because [A] is falling.

Example 8.1 Find the rate of reaction at 2.5 and 5.0 minutes, using the data in Table 8.1.
If [A] is plotted against time, the Graph 8.1 is obtained.

You can see that it is not easy to get accurate values for the rates of reaction at 2.5 and 5.0 minutes because these are the gradients at these times, and the graph is a curve rather than a straight line. This implies that the rate of reaction is constantly changing, and this is of course because the concentration of A, which determines the rate, is constantly falling. One way of

Table 8.1 Data for Example 8.1.

Time (minutes)	[A] *(mmol/l)*
0	80
1	52
2	33
3	22
4	14
5	9.5
6	6.0
7	4.1
8	2.6
9	1.6

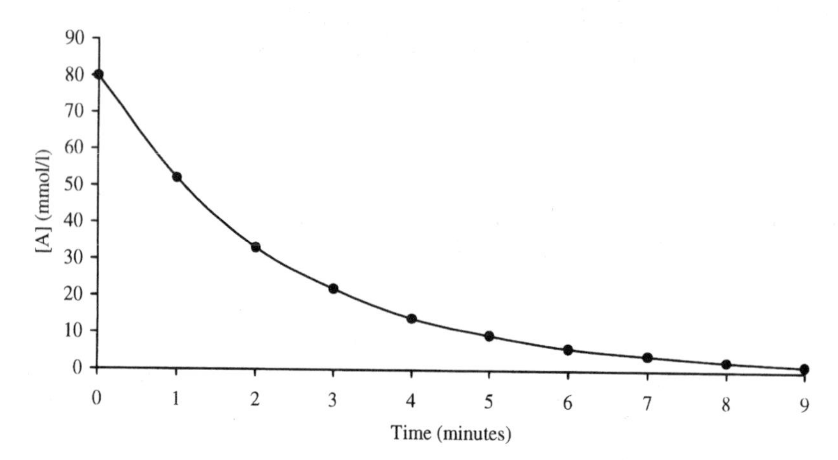

Graph 8.1

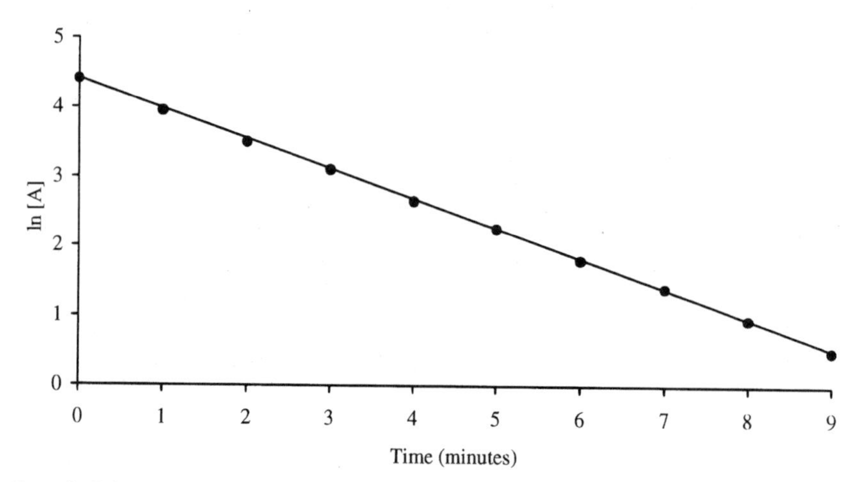

Graph 8.2

calculating the gradient more easily is to transform the data mathematically so that when replotted the graph becomes a straight line. In this example you calculate the natural logs of the values for $[A]$. Graph 8.2 shows a plot of ln[A] against time.

You can see that this is a straight line. Therefore its equation must be of the form $y = ax + b$. In this case, $y = \ln[A], x =$ time, and b is the y-intercept which $=$ 4.38. The gradient is a, which you can calculate in the usual way from two points, and which equals –0.429. Therefore the equation of this straight line is:

$$\ln[A] = -0.429\ (\text{time}) + 4.38$$

The gradient a is also equal to the rate constant, k. The rates of reaction at 2.5 and 5.0 minutes can be obtained by first finding $[A]$ at these times.

So, at 2.5 minutes,
$\ln\ [A] = -0.429(2.5) + 4.38 = 3.31$
So $[A] = \text{antiln}(3.31) = 27.3\,\text{mmol/l}$
and since rate of reaction $= \text{k}[A]$
rate of reaction $= -0.429(27.3)$
$= -11.7\ \text{mmol}\,\text{l}^{-1}\ \text{minute}^{-1}$.

Similarly, at 5 minutes,
$\ln[A] = -0.429(5) + 4.38 = 2.26$
so $[A] = \text{antiln}(2.26) = 9.54\,\text{mmol/l}$
and rate of reaction $= -0.429(9.54)$
$= -4.09\ \text{mmol}\ \text{l}^{-1}\ \text{minute}^{-1}$.

Note that the rates of reaction are negative because they represent the rate of fall in concentration of A.

The example above shows how it may be possible to obtain an equation relating two variables if one or both variables can be transformed so that when the transformed data is plotted, a straight line is obtained. Therefore you might expect that several shapes of curve can be described by relatively simple equations. Graphs 8.3 to 8.8 are examples.

Graph 8.3 is of the equation $y = be^{ax}$ where both a and b are positive (in this case, 0.5 and 2 respectively). This is an exponential equation.
Graph 8.4 is of the equation $y = be^{ax}$ where a is negative and b is positive (in this case, –0.5 and 5, respectively). This is also an exponential equation.
Graph 8.5 is of the equation $y = a(\ln x)$ where a is positive (in this case 0.5). This is a logarithmic equation.
Graph 8.6 is of the equation $y = bx^a$ where both a and b are positive (in this case, 0.5 and 2 respectively).
Graph 8.7 is of the equation $y = ax^2$ where a is positive (in this case 0.5).
Graph 8.8 is of the equation $y = {}^{ax}/_{(b+x)}$ where a and b are positive (in this case, 0.75 and 1.25 respectively).

In each of these cases, it is possible to obtain an equation relating x and y, and the values of the constants a and b, by transforming the data in the following ways:

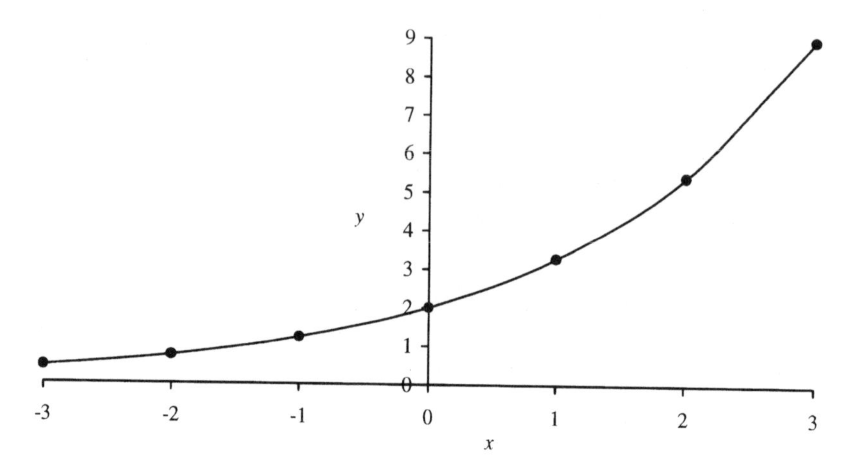

Graph 8.3

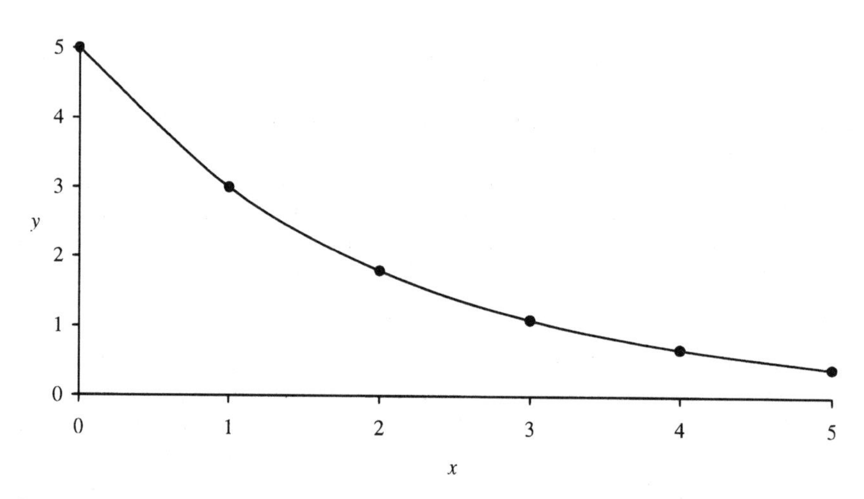

Graph 8.4

- For equations of the form $y = be^{ax}$ (Graphs 8.3 and 8.4)

$$\ln y = \ln(be^{ax}) = \ln \mathrm{b}+\ln(\mathrm{e}^{ax})$$
$$= \ln b + ax$$

If $\ln y$ is plotted against x, the gradient $= a$ and the intercept on the $\ln y$ axis $= \ln b$.

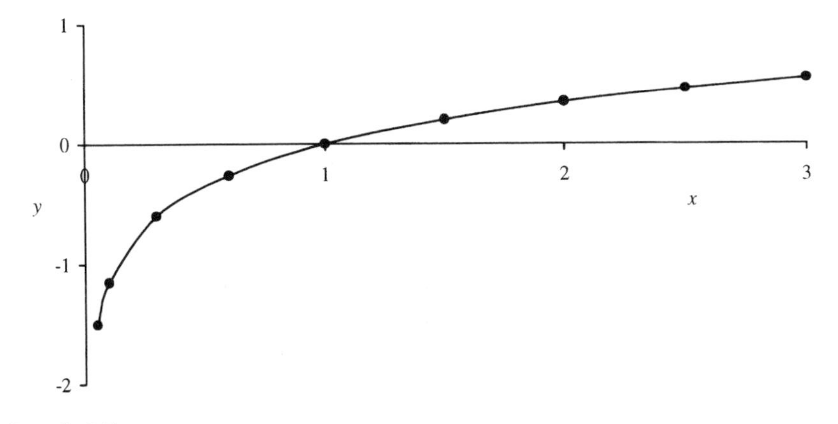

Graph 8.5

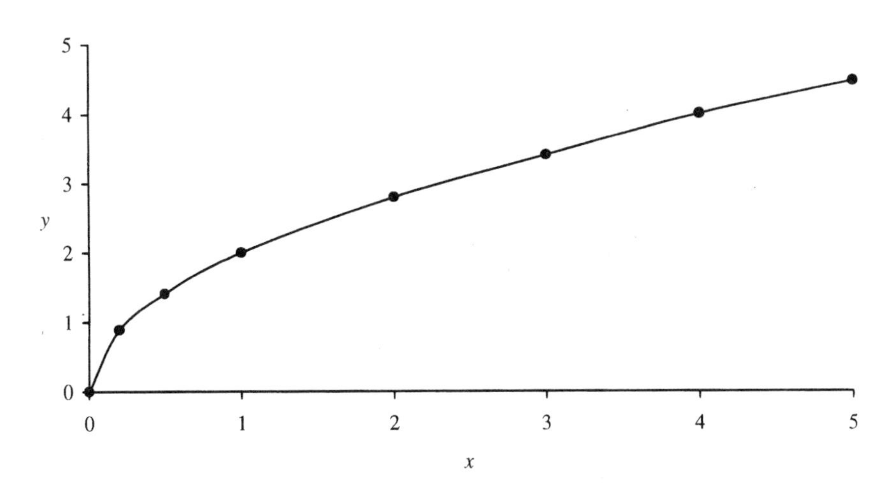

Graph 8.6

- For equations of the form $y = a(\ln x)$ (Graph 8.5)

If y is plotted against $\ln x$, the gradient $= a$.

- For equations of the form $y = bx^a$ (Graph 8.6)

$$\begin{aligned} \ln \mathrm{y} &= \ln(bx^a) = \ln b + \ln x^a \\ &= \ln b + a(\ln x) \end{aligned}$$

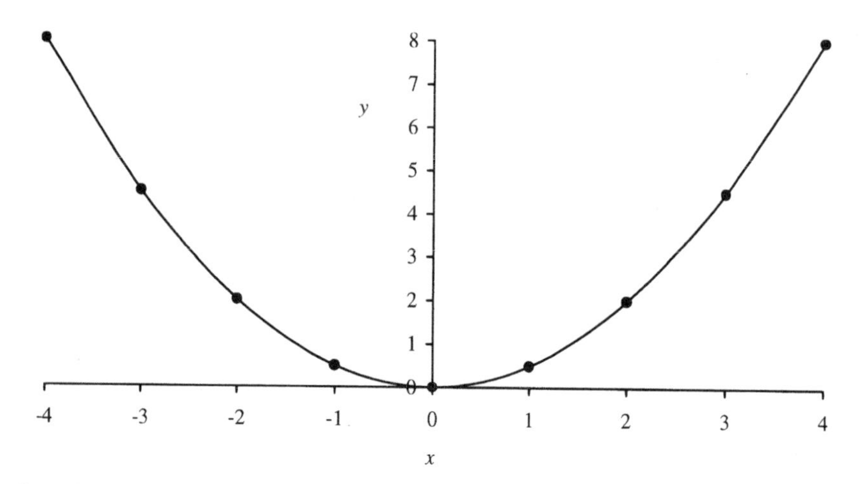

Graph 8.7

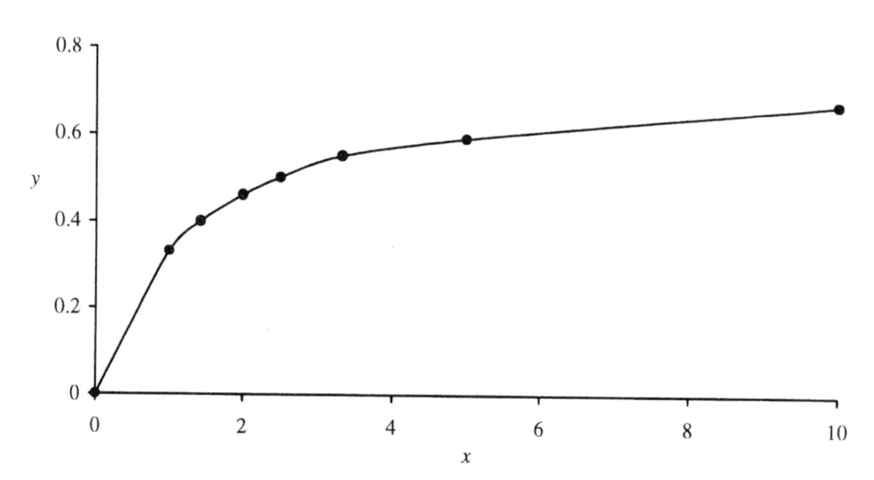

Graph 8.8

If $\ln y$ is plotted against $\ln x$, the gradient $= a$ and the intercept on the $\ln y$ axis is $\ln b$.

- For equations of the form $y = ax^2$ (Graph 8.7)

If y is plotted against x^2, the gradient $= a$.

- For equations of the form $y = {}^{ax}/_{(b+x)}$ (Graph 8.8) there are several possibilities.

(i) Invert the equation :

$$\frac{1}{y} = \frac{b+x}{ax}$$

This is the same as

$$\frac{1}{y} = \frac{b}{a} \times \frac{1}{x} + \frac{1}{a}$$

If $^1/_y$ is plotted against $^1/_x$, the gradient $= {}^b/_a$, and the $^1/_y$ intercept $= {}^1/_a$. Also, the $^1/_x$-intercept $= -{}^1/_b$.

(ii) Invert the equation :

$$\frac{1}{y} = \frac{b+x}{ax}$$

Multiply by x

$$\frac{x}{y} = \frac{b+x}{a} = \frac{b}{a} + \frac{1}{a} \times x$$

If $^x/_y$ is plotted against x, the gradient is $^1/_a$ and the $^x/_y$ intercept is $^b/_a$.

(iii) Multiply by the denominator, then divide by x:

$$y = \frac{ax}{b+x}$$

$$by + xy = ax$$

$$\frac{by}{x} + y = a$$

$$y = a - \frac{by}{x}$$

If y is plotted against $^y/_x$, the gradient is $-b$ and the y-intercept is a.

8.2 RATES OF CHANGE

In the previous section you have seen how variables may be related in such a way that the rate of change of y with a change in x is itself constantly changing. In example 8.1 where the rate of change was the rate of a reaction = the rate of disappearance of A, I introduced the expression $-d[A]/dt$

where the 'd' stands for 'a small change in', and the minus sign is present because [A] is falling. The term $^{dy}/_{dx}$ is a general term for the change in y relative to the change in x and is therefore the gradient of the graph of y against x for any shape of graph. The value of $^{dy}/_{dx}$ at any particular value of x is given by the 'differential form' of the equation of the graph of y against x. The latter can be called the 'integrated form' of the equation, and I gave examples of these in Section 8.1. (An explanation of the derivation of the differentiated forms is outside the scope of this book. You would need to read an 'A' level or higher level textbook to find out about it. However, you can use the differentiated forms of the equations in calculations without understanding their origins.) The differential forms of the equations are listed in Table 8.2.

By using these equations you can therefore calculate $^{dy}/_{dx}$, the rate of change of y with x, for any data whose points form a line which can be described by one of the listed integrated forms.

Given a set of data values of x and y, you should do the following.

(i) Plot y against x. If this gives an obvious straight line, you can determine the gradient and the intercept directly, and hence find $^{dy}/_{dx}$, the equation of the line, and therefore the relationship between y and x.
(ii) If the plot does not give an obvious straight line, see whether it more closely resembles any of the types of curve discussed above. If so, try transforming the data appropriately and replot. If you now obtain a straight line, you can determine the gradient and intercept, find the equation of the line, and find $^{dy}/_{dx}$ from the differentiated form of the equation.

Example 8.2 The number of bacteria in a culture was measured at 10 minute intervals over one hour. The results are shown in Table 8.3.

(i) What is the relationship between number of bacteria and time?
(ii) Calculate the growth rate at 15 minutes.

Table 8.2 Integrated and Differentiated Forms of Equations.

Integrated form	*Differentiated form*
$y = ax + b$	$^{dy}/_{dx} = a$
$\ln y = ax + b$	$^{dy}/_{dx} = ay$
$\ln y = ax + \ln b$ or $y = be^{ax}$	$^{dy}/_{dx} = ay$
$y = a(\ln x)$	$^{dy}/_{dx} = {}^{a}/_{x}$
$y = ax^2$	$^{dy}/_{dx} = 2ax$
$y = bx^{a}$	$^{dy}/_{dx} = bax^{a-1}$

Table 8.3 Data for Example 8.2.

Time (minutes)	*No. of bacteria/ml*
0	3.0×10^4
10	4.2×10^4
20	6.1×10^4
30	8.5×10^4
40	11.9×10^4
50	16.9×10^4
60	23.8×10^4

A plot of number of bacteria/ml against time gives an upward curving plot that is obviously not a straight line, but suggests an exponential relationship, i.e. if the number of bacteria/ml at time $t = N_t$, then
$N_t = be^{at}$ where a and b are constants.
A plot of ln N_t against t gives a straight line with a gradient of 0.0345 min^{-1} and intercept of 10.3.
Therefore $\ln N_t = 10.3 + 0.0345t$
$\text{antiln}(\ln N_t) = \text{antiln}(10.3 + 0.0345t)$
$= \text{antiln}(10.3) \times \text{antiln}(0.0345t)$
$N_t = 3.01 \times 10^4 \times e^{0.0345t}$
Also, since $dN/dt = aN_t$
at time 15 minutes,
$N_t = 3.01 \times 10^4 \times e^{0.0345(15)}$
$= 3.01 \times 10^4 \times e^{0.518}$
$= 3.01 \times 10^4 \times 1.68$
$= 5.05 \times 10^4$
so growth rate $= dN/dt = 0.0345 \times N_t$
$= 0.0345 \times 5.05 \times 10^4$
$= 1742$ bacteria/minute at 15 minutes

Example 8.3 The rate of a reaction was measured at different temperatures. From the results shown in Table 8.4, determine the relationship between rate of reaction and temperature.

Table 8.4 Data for Example 8.3.

Rate (μmol/minute)	*Temperature (K)*
24.7	283
34.3	298
47.3	310
57.3	318
70.0	328

Graph 8.9 shows the rate plotted against the temperature. The points suggest an upward curve rather than a straight line. This suggests y and x are likely to be related according to an equation of the form $y = be^{ax}$, and therefore you should try plotting ln rate against temperature. This is shown in Graph 8.10. You can see the points appear to lie on a straight line with gradient 0.024, suggesting the equation:

$$\begin{aligned} \ln \text{rate} &= 0.024\,(\text{temperature}) + \ln b, \\ \text{or rate} &= be^{0.024(\text{temp})} \end{aligned}$$

However, if you plot ln rate against $^1/_{\text{temperature}}$ you obtain Graph 8.11. Again, the points appear to fit a straight line, this time with a gradient of -2.15×10^3, suggesting the equation:

$$\begin{aligned} \ln \text{rate} &= -2.15 \times 10^3 (^1/_{\text{temp}}) + \ln b, \\ \text{or rate} &= be^{-2150/\text{temp}} \end{aligned}$$

This illustrates a potential problem in that more than one transformation may give an apparently satisfactory fit of a curve to the data. You need to be aware of this possibility. It may be possible to eliminate one equation by thinking about what values are obtained at the extremes. In the example, using the first equation, when temperature is 0 K, rate $= b$ which is non-zero.

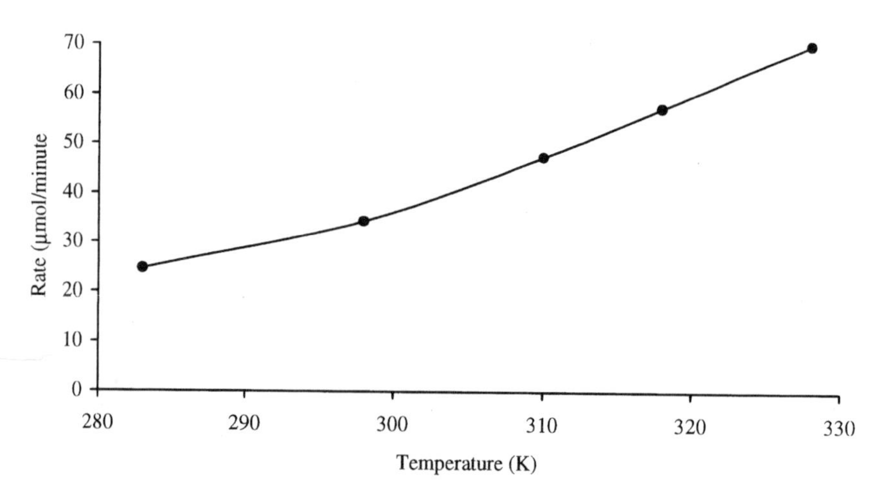

Graph 8.9

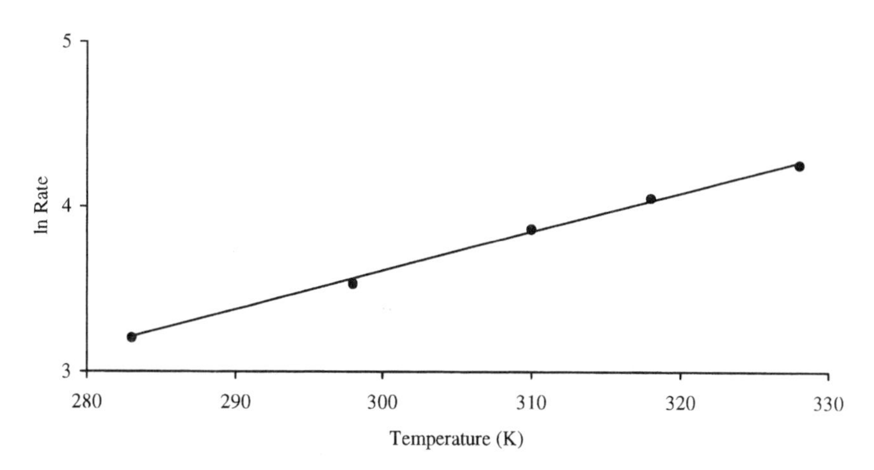

Graph 8.10

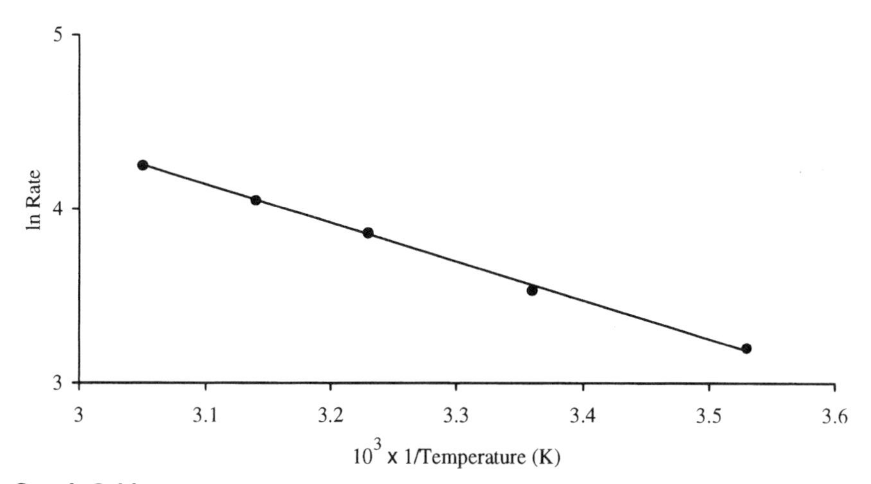

Graph 8.11

This cannot be true, though, since at absolute zero the molecules would be motionless and could not react. Using the second equation, you obtain rate = 0 at 0 K, which is sensible.

Transforming data into a logarithmic form is also often useful when the values for the independent variable span several orders of magnitude. By taking the log of these values it is easier to plot a graph which shows the effective range of values over which the dependent variable changes significantly. This is particularly applicable in physiology or pharmacology where you may be interested in the relationship between the concentration of a neurotransmitter or the dose of a drug and the response of a tissue.

8.3 PROBLEMS

If you have just finished reading this chapter I suggest you now try the 'pure mathematics' examples marked'*'. If you get these correct, try *all* of the applied examples. If you also get these correct, congratulations! If you do not get the right answers, re–read the explanations and try the rest of the problems 1–9, and then repeat your attempts at the applied problems.

Sketch graphs of the following equations and of their differentiated forms. Find the values of ${}^{dy}/_{dx}$ at $x = 2$.

1. $y = e^x$	2. $y = e^{x/3}$	**3.*** $y = 2e^{3/x}$
4. $y = e^{-x}$	5. $y = \ln x$	**6.*** $y = 10(\ln x)$
7. $y = \ln x^2$	**8.*** $y = \ln\left({}^{1}/_{x}\right)$	9. $y = 5x^{0.5}$

10. In an experiment to measure the effect of histamine on the contraction of isolated guinea-pig ileum, the extent of tissue contraction was measured at 11 different concentrations of histamine. The results are shown in Table 8.5. Calculate the histamine concentration which gives half the maximum effect.

Table 8.5 Data for Question 10, Chapter 8.

Histamine concentration	*Tissue contraction (mm)*
50 nmol/l	0.0
100 nmol/l	0.0
200 nmol/l	0.2
400 nmol/l	0.7
800 nmol/l	2.1
1.6 μmol/l	4.6
3.2 μmol/l	7.6
6.4 μmol/l	9.8
12.8 μmol/l	11.3
25.6 μmol/l	12.2
51.2 μmol/l	13.4
102.4 μmol/l	13.4
204.8 μmol/l	13.4

11. A hormone H binds reversibly to its receptor R on cell membranes:

$$H + R \leftrightarrow HR$$

The equilibrium constant for the dissociation of hormone-receptor complex into free hormone and receptor is K_d. The data in Table 8.6 gives values for the amount of hormone bound by receptor at different hormone concentrations. Calculate the total number of receptors and the K_d.

12. Radioactivity arises from atoms which have unstable nuclei. These nuclei disintegrate spontaneously, emitting ionising radiation. Radioactivity is quantified by the rate of disintegration. For example, a sample of a radioactive substance can be said to have an activity of 5×10^6 disintegrations per second, meaning that in the sample, this number of nuclei disintegrate every second. Obviously, if nuclei are disintegrating, the activity of a sample must decrease over time as there are fewer nuclei left to disintegrate. The time it takes for the activity to fall to half of its original activity is called the half-life, $t_{1/2}$. This is a characteristic of each radioisotope, i.e. each type of radioactive atom. Calculate the half life of a radioisotope X, and the activity after 2 weeks from time 0, given the data in Table 8.7.
13. For a simple reaction X<-->Y the rate of reaction $= k[X]$ where k is called the rate constant. The value of k depends on the temperature and the relationship is given by $k = Ae^{-Ea/RT}$ where A is a constant depending on the nature of the reaction, R is the gas constant (= 8.3 joules $mol^{-1}K^{-1}$) and Ea is called the activation energy of the reaction. T is the absolute temperature (K). Calculate Ea for a reaction, given the values of the rates of reaction at different temperatures shown in Table 8.8.
14. The data in Table 8.9 relate the average basal metabolic rates of several species of mammal to the average body masses of these mammals. Derive an equation that describes the apparent relationship between these two variables.
15. Using the data from question 20 in Chapter 7, find the relationship between rate and substrate concentration, and the value of the constants, by using three different transformations of the equation:

$$\text{rate} = {}^{\text{a[substrate]}}\!/\!_{(\text{b} + [\text{substrate}])}.$$

Table 8.6 Data for Question 11, Chapter 8.

[Bound hormone] (fmol/ml)	*[Free hormone] (nmol/l)*
9.7	10.0
9.3	8.0
9.0	6.0
8.5	4.0
7.0	2.0
5.2	1.0
3.5	0.5

Table 8.7 Data for Question 12, Chapter 8.

Activity (disintegrations/s)	*Time (days)*
2.40×10^8	0
1.56×10^8	1
0.99×10^8	2
0.66×10^8	3
0.42×10^8	4
0.26×10^8	5
0.18×10^8	6

Table 8.8 Data for Question 13, Chapter 8.

Rate (μmol/min)	$T(K)$
7.4	283
10.3	298
14.2	310
17.2	318
21.0	328

Table 8.9 Data for Question 14, Chapter 8.

Species	*Metabolic rate (watts)*	*Body mass (kg)*
Cattle	340	350
Sheep	55	28
Dogs	25	15
Cats	7.3	2.9
Rats	1.3	0.25
Mice	0.33	0.054
Rabbits	10.0	4.2
Guinea pigs	3.1	0.92
Hamsters	0.9	0.11

ANSWERS TO PROBLEMS

CHAPTER 2

1. 1
2. $^{5}/_{16}$
3. $^{108}/_{105}$ or $1\,^{1}/_{35}$
4. $^{1}/_{2}$
5. $^{13}/_{12}$ or $1\,^{1}/_{12}$
6. $^{27}/_{35}$
7. $^{14}/_{9}$ or $1^{5}/_{9}$
8. $^{25}/_{18}$ or $1^{7}/_{18}$
9. $^{81}/_{28}$ or $2\,^{25}/_{28}$
10. $^{6}/_{5}$ or $1^{1}/_{5}$
11. 30.66
12. 4.14
13. 1.04
14. 26.95
15. 169.638
16. 1.225
17. 3.45
18. 13.16
19. 5.45
20. 23.75
21. 0.63
22. 6.9
23. 3.51
24. 33.3%
25. 75%
26. 77.8%
27. 88.3%
28. 60
29. 28
30. 80
31. 25.1%
32. 0.45 g of protein; 0.21 g of nucleic acid; 0.24 g
33. See Table 9.1
34. 21.8%; $^{28}/_{100}$; 0.28
35. (i) 3 g of sucrose made up to 100 ml with solvent
    (ii) 15 g of sucrose made up to 500 ml

Table 9.1 Answer to Question 32, Chapter 2.

Element	*Mass*	%
H	6.51	9.3
C	13.65	19.5
N	3.57	5.1
O	43.96	62.8
Ca	0.98	1.4
P	0.43	0.61
S	0.45	0.64
Na	0.18	0.26
K	0.15	0.21
Cl	0.13	0.18

(iii) 25 g of sucrose plus 475 g solvent
(iv) 2.5 ml of ethanol plus 497.5 ml solvent.
(v) 2 g of ethanol made up to 20 ml with solvent

36. (i) 0.15 g
(ii) 0.15 g
(iii) 0.25 g

CHAPTER 3

1. $^2/_1$
2. $^8/_5$ or $1\,^3/_5$
3. $^1/_8$
4. $^1/_4$
5. $^{3g}/_{7g}$
6. 30 ml/l or 0.03 l/l
7. 12
8. $^2/_3$
9. 32
10. 49
11. 2
12. $^{2.5}/_3$ or 0.833
13. 50 000 cm or 500 m or 0.5 km
14. 1:500 000
15. 4.541
16. 400
17. See Table 9.2
18. (i) 275.5 g (ii) 275.5 g
(iii) 27.55 g (iv) 27.55%(w/v)
(v) 0.0374%(w/v) (vi) 0.00226 mol/l
(vii) 0.181 l or 181 ml (viii) 0.73 ml
(ix) and (x) See Table 9.3
19. Using 0.5 mol/l stock solution and water,
(i) 0.4 ml stock + 9.6 ml water
(ii) 0.08 ml stock + 9.92 ml water, or, 1 ml stock + 4 ml water, then 0.4 ml of this + 9.6 ml water.

Table 9.2 Answers to Question 17, Chapter 3.

Volume of 5% (w/v) solution	*Mass in given volume*	*Moles of Na^+ in given volume*
100 ml	5 g	0.085
1 l	50 g	0.85
5 ml	0.25 g	0.00425
0.01 ml	0.5 mg or 0.0005 g	0.0000085
20 l	1000 g or 1 kg	17

Table 9.3 Answers for Questions 17 (ix) and (x), Chapter 3.

Volume of 0.01 mol/l ATP (ml)	*Volume of water (ml)*	*Concentration (mol/l)*	*Concentration (mol/l) in part (x) of question*
5	5	0.005	0.000125
4	6	0.004	0.000100
3	7	0.003	0.000075
2	8	0.002	0.000050
1	9	0.001	0.000025
0	10	0.000	0.000000

(iii) 1 ml stock +4 ml water, then 0.1 ml of this + 0.9 ml water.
(iv) 1 ml stock +4 ml water, then 0.3 ml of this + 0.7 ml water.
(v) 0.11 ml stock + 9.89 ml water.
(vi) 1 ml stock + 8 ml water, then 0.15 ml of this + 4.85 ml water.
(vii) 1 ml stock + 17 ml water, then of 0.08 ml of this + 1.92 ml of water.
(viii) 1 ml of stock + 17 ml of water, then 0.3 ml of this + 0.7 ml of water.

The answers given below for (ix) and (x) are not the only possible correct answers. Generally you should be trying to be practical and economical.

(ix) 0.4 ml of stock + 9.6 ml of water; then of this, 1.0, 0.75, 0.50, 0.25 and 0 ml with 0, 0.25, 0.50, 0.75 and 1.0 ml water respectively.
(x) 1 ml stock + 44 ml water; then of this, 10, 7.5, 5.0, 2.5 and 0 ml with 0, 2.5, 5.0, 7.5 and 10 ml of water respectively.

20. 120 ml/minute

CHAPTER 4

1. 8
2. 1
3. $^1/_{27}$ or 0.037
4. 0.01 or $^1/_{100}$
5. $^{81}/_{10000}$ or 0.0081
6. 2
7. $^4/_9$ or 0.444
8. $^4/_9$ or 0.444
9. $^{625}/_{256}$ or 2.44
10. a^5
11. a
12. a^{-5} or $^1/_{a^5}$

13. a^2
14. a^{-5} or $1/a^5$
15. a^{-1} or $1/a$
16. a^6
17. a^{-2} or $1/a^2$
18. a^6
19. 10
20. 0.1
21. 0.1
22. 3.5×10^{-5}
23. 1.73×10^{10}
24. 3.184576×10^3
25. 3.73×10^{-4}
26. 1.6×10^2
27. 3.51×10^{-4}
28. 0.000315
29. 0.000376
30. 150
31. 0.015
32. 0.04786
33. 61
34. 1.093×10^4
35. 10^9
36. 2.89×10^4
37. -3.04×10^{-2}
38. 4.495×10^{-2}
39. -4.35×10^{-2}
40. 270 or 2.7×10^2
41. 7.5×10^4
42. 3.96×10^{-2}
43. 3×10^{-8}
44. 3×10^{-2}
45. 0.3
46. 0.471 g
47. 0.018 1
48. 140 μl
49. 51 μg
50. 0.042 mg
51. 0.831 ml
52. 1.4 μl
53. 37 μmol/l
54. 0.031 mmol/1
55. 17 mol/m^3
56. 3.1 μmol/1
57. 1.4×10^{-5} mmol/ml
58. 1.5×10^3 mg/1
59. 3.1×10^{-5} % (w/v)
60. See Table 9.4
61. Total phosphate is 0.092%; 9.6 mmol/l.

    Chloride is 0.50%; 142 mmol/l.
62. Na^+ is 0.35%; 153 mmol/l.

    K^+ is 0.016%; 4.18 mmol/l.

    Mg^{2+} is 1.2×10^{-3} %; 0.49 mmol/1.

    Ca^{2+} is 3.5×10^{-3} %; 0.88 mmol/1.

Table 9.4 Answer for Question 60, Chapter 4.

Components	*Formula weight*	*Concentration % (w/v)*	*Concentration mmol/l*
Na_2HPO_4	142	0.115	8.1
KH_2PO_4	136	0.02	1.5
NaCl	58.5	0.80	137
KCl	74.5	0.02	2.68
$MgCl_2 . 6H_2O$	203	0.01	0.49
$CaCl_2 . 2H_2O$	147	0.013	0.88

Table 9.5 Answers to Question 63, Chapter 4.

Volume of 50 mmol/l solution	*Volume of water added*	*Concentration*	*Weight of pyruvic acid in 0.2 ml of this solution*
10 ml	90 ml	5 mmol/l	0.11 mg
0.5 ml	19.5 ml	1.25 mmol/l	27.5 μg
100 μl	0.9 ml	5 mmol/l	110 μg
25 μl	475 μl	2.5 mmol/l	55 μg
1.5×10^{-2} l	3.5×10^{4} ml	0.021 mmol/l	0.47 μg
20 ml	20 ml	25 mmol/l	0.55 mg
10 μl	490 μl	1.0 mmol/l	2.2×10^{-5} g
70 μl	100 ml	35 μmol/l	7.7×10^{-4} mg

63. 0.55 g. See Table 9.5 for other answers.

64. 2.3×10^{7} bacteria/ml.

CHAPTER 5

1. −3
2. −2
3. 5.5
4. 2
5. 3
6. 2
7. 4.8
8. 0.5
9. 4.167
10. −5.96
11. 39.3
12. 54.47
13. −19.5
14. 14.67
15. 6
16. 5
17. 7
18. 1.4
19. −2.57
20. −3.8
21. 3 and −1
22. −2 and 1
23. 3 and −2
24. 5.284 and −0.284
25. 6.15 μmol/l
26. (i) 0.1 μmol/sec
    (ii) 0.83
    (iii) 0.8 mmol/1
    (iv) 4.0 mmol/1
    (v) See Table 9.6
27. 2.99×10^{-3} mol/1

Table 9.6 Answers to Question 26 (v), Chapter 5.

$[S]/K_m$	v/V_{max}
0.01	0.0099
0.10	0.091
0.50	0.333
1.0	0.50
3.0	0.75
5.0	0.833
10.0	0.91
100	0.99

CHAPTER 6

1. 4
2. 9
3. 2
4. 4
5. 0
6. 5
7. −4
8. −3
9. −7
10. 6
11. 4
12. 3
13. −2
14. $^5/_3$
15. $^1/_2$
16. $^3/_5$
17. 0.60
18. 1.56
19. −0.22
20. 0.18
21. −0.18
22. 0.70
23. 2.30
24. 4.48
25. −2.22
26. −1.40
27. 2
28. 0.566
29. −6
30. 1.098
31. 16
32. 81
33. 128
34. 0.008
35. 0.0625
36. 2
37. 8
38. 10^9
39. 1.0023
40. 0.977
41. 2
42. 13
43. 0.125
44. 5
45. 0.02
46. 0.0126
47. 1999
48. 4002
49. 31.64
50. 3.0000004
51. 5 or −2
52. 1
53. $\ln y = \ln a + b \ln x$
54. $y = \text{antilog} b \times x^a$
55. (i) See Table 9.7
    (ii) 31.6%
56. (i) 1.0 (ii) 4 (iii) 2.3
    (iv) 10.1 (v) 2.22
57. (i) 13.0 (ii) 11.0 (iii) 11.48
    (iv) 9.23 (v) 7.04

Table 9.7 Answers to Question 55, Chapter 6.

Percentage T	*A*
100	0
10	1
1.0	2
0.1	3

Table 9.8 Answers to Question 58, Chapter 6.

pH	$[H^+]$	$[OH^-]$
2.1	7.9×10^{-3} mol/l	1.3×10^{-12} mol/l
8.0	1.0×10^{-8} mol/l	1.0×10^{-6} mol/l
3.6	2.5×10^{-4} mol/l	4.0×10^{-11} mol/l
11.4	4.0×10^{-12} mol/l	2.5×10^{-3} mol/l

58. See Table 9.8
59. 10.6
60. Ethylamine gives pH 11.4 for a 0.01 mol/1 solution.
61. 4.34
62. Carboxyl group 99.999% ionised
    Amino group 99.5% ionised
63. $H_2PO_4^- = 0.22$ mmol/1
    $HPO_4^{2-} = 0.78$ mmol/1
    $PO_4^{3-} = 7.8$ nmol/1
64. $\Delta G^\circ = 1783$ J/mol

CHAPTER 7

1. 1.
2. 2.25.
3. 1.67
4. −1.467
5. $y = x - 1$.
6. $y = 2.5x - 6.5$
7. $y = -0.25x + 3.5$
8. $y = 3x + 2$
9. $y = 2x + 4$
10. $y = -x - 4$
11. $y = 0.25x + 4.25$
12. $y = -0.6x$

13–16. See Graphs 9.1–9.4

17. See Graph 9.5. Gradient = E = 18.2 l $mmol^{-1}$ cm^{-1}
    = 18.2×10^3 l mol^{-1} cm^{-1}.

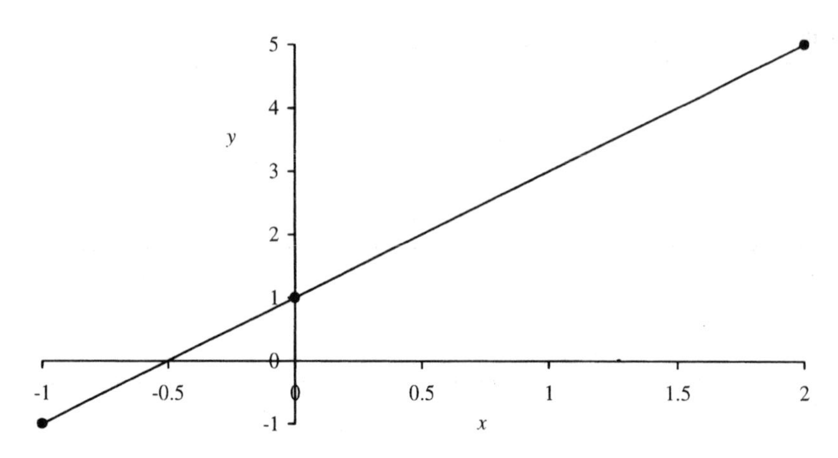

Graph 9.1

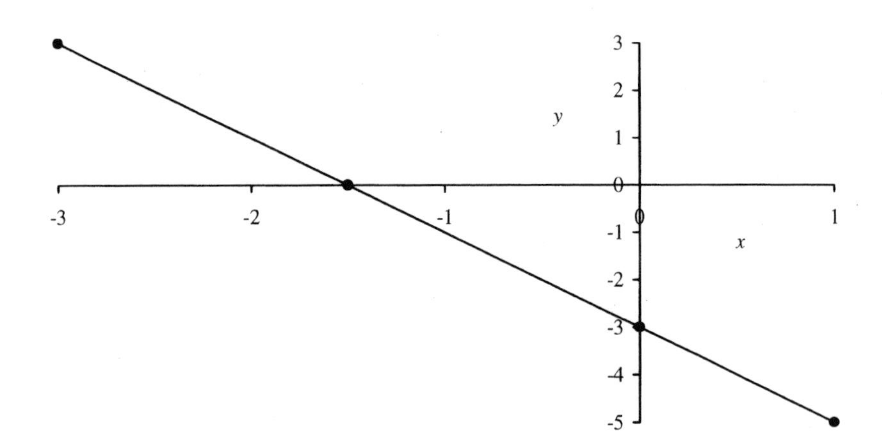

Graph 9.2

18. See Table 9.9
19. (i) 0.02 absorbance minute^{-1}
    (ii) 0.0267 μmol minute^{-1}
    (iii) 0.267 μmol minute^{-1} ml^{-1} saliva
    (iv) 0.1 μmol minute^{-1} ml^{-1} saliva

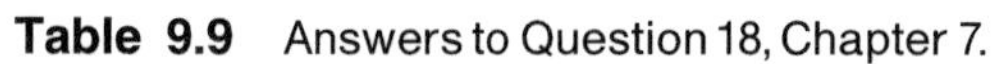

Table 9.9 Answers to Question 18, Chapter 7.

Unknown	*Volume of unknown (ml)*	*Volume of buffer (ml)*	*Absorbance*	*Concentration*
A	5.0	0.0	0.45	0.025 mmol/l
B	1.0	4.0	0.30	0.082 mmol/l
C	0.5	4.5	0.65	0.357 mmol/l
D	0.3	2.7	0.50	0.275 mmol/l

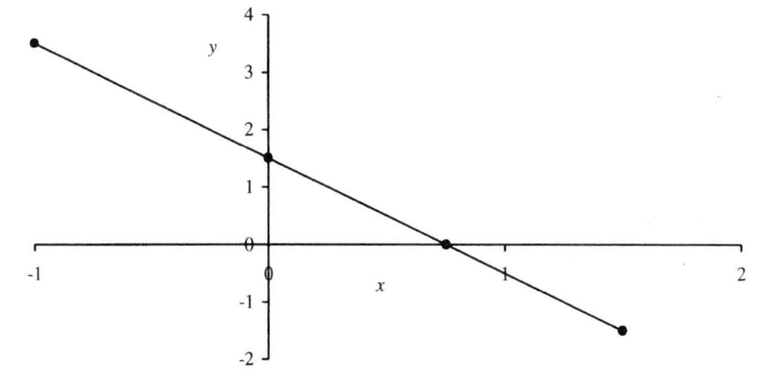

Graph 9.3

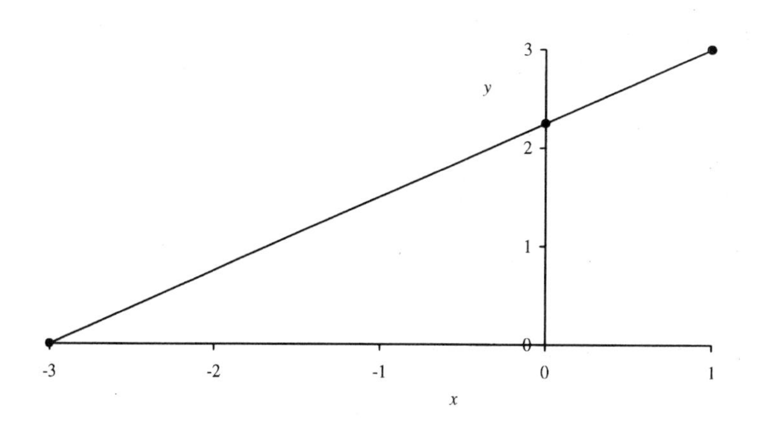

Graph 9.4

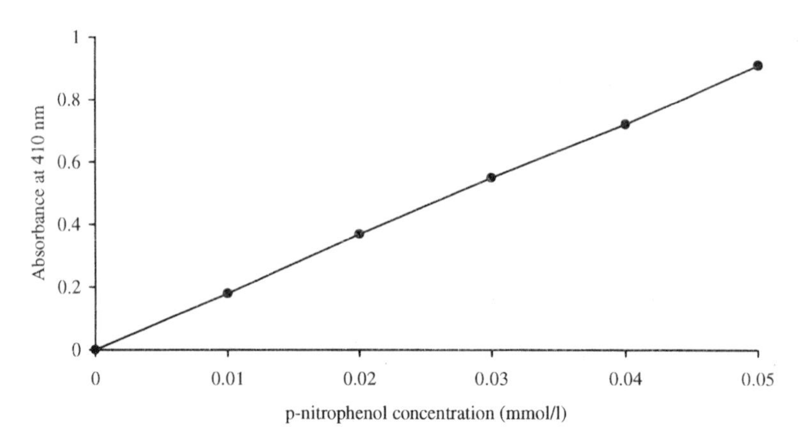

Graph 9.5

20. The graph is *not a straight line*, though it approximates to one when the substrate concentration is low. Therefore the rate of reaction is *not* proportional to the substrate concentration, though it approaches proportionality at low substrate concentrations.

CHAPTER 8

1–9. See Graphs 9.6–9.23

10. Plot mm contraction vs log [histamine]. The result is an S-shaped curve. Since the maximum contraction is 13.4 mm, the half-maximum effect is 6.7 mm, and at this point the log [histamine] is –5.59, so the [histamine] is 2.6 µmol/1.
11. Plot [bound hormone] vs [bound]/[free hormone].
    Gradient $= -1\text{nmol/l} = K_d$
    The bound-intercept = 10.5 fmol/ml = total number of receptors/ml
    so 10.5 fmol/ml $\times\ 6.02 \times 10^{23} = 6.3 \times 10^9$ receptors/ml.
12. Plot ln dps vs time.
    Gradient = –0.433, y-intercept = 19.3
    therefore ln dps = –0.433(time) + 19.3
    (i) For t = half-life, ln dps = ln $(2.4 \times 10^8/2) = 18.6$
    therefore $18.6 = -0.433(t_{1/2}) + 19.3$
    $t_{1/2} = (19.3 - 18.6)/0.433 = 1.61$ days

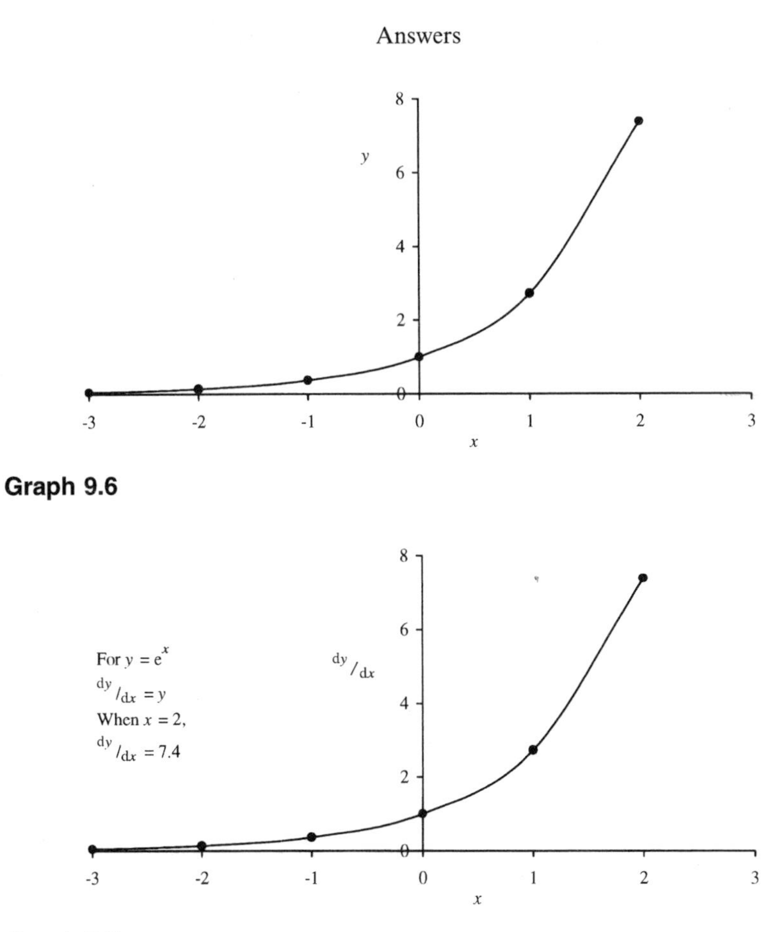

Graph 9.6

Graph 9.7

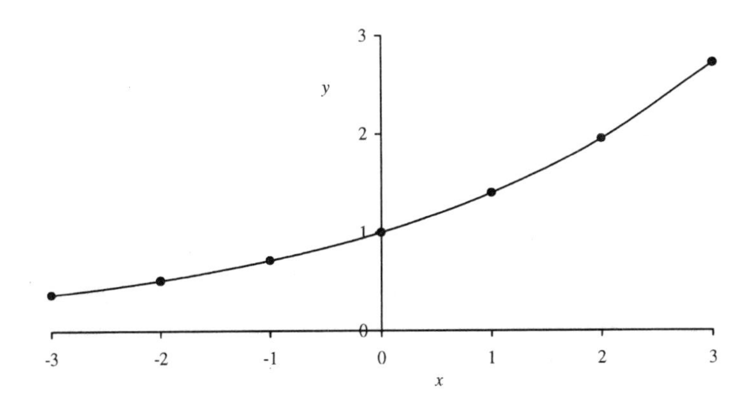

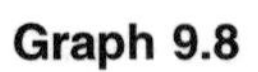

Graph 9.8

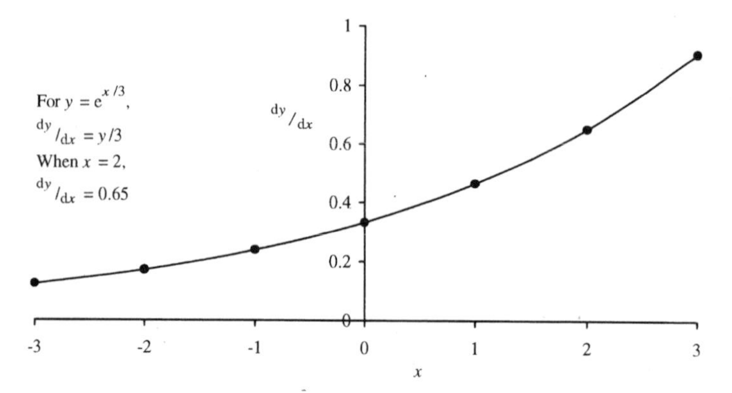

Graph 9.9

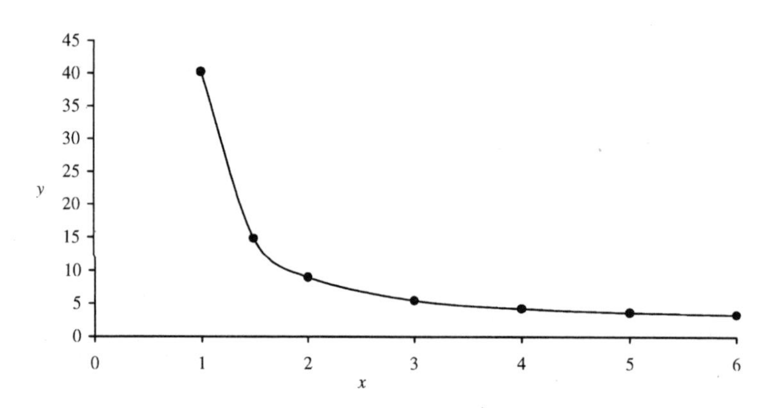

Graph 9.10

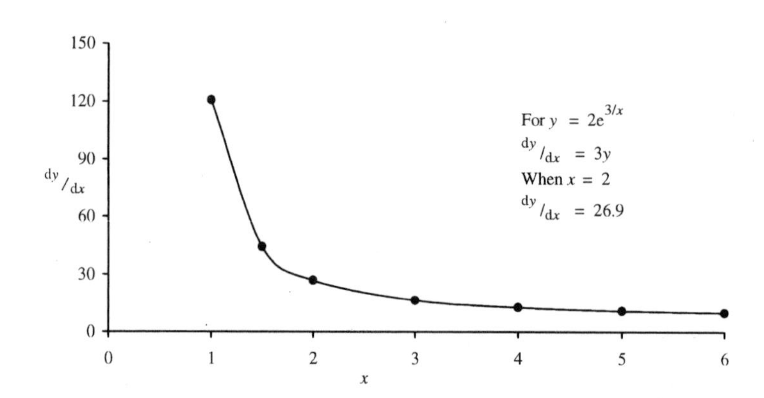

Graph 9.11

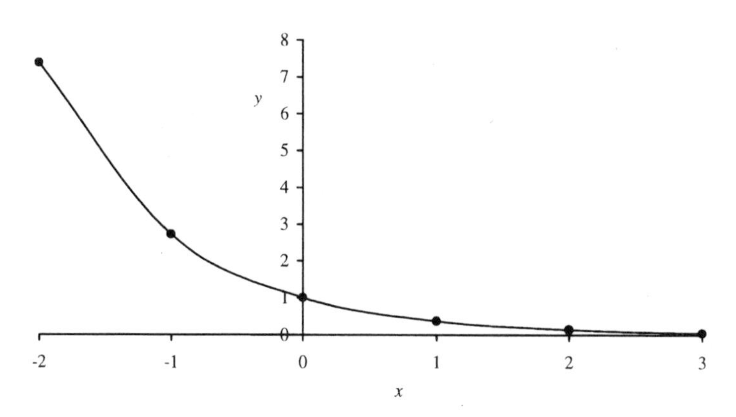

Graph 9.12

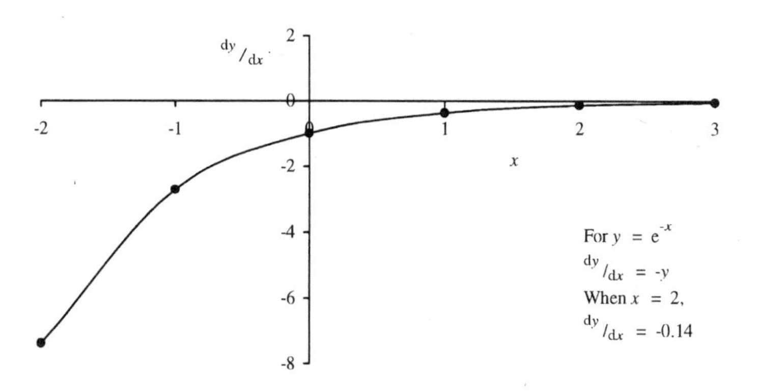

Graph 9.13

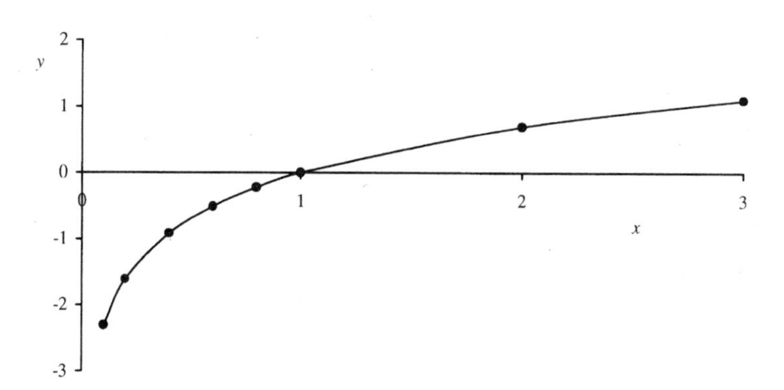

Graph 9.14

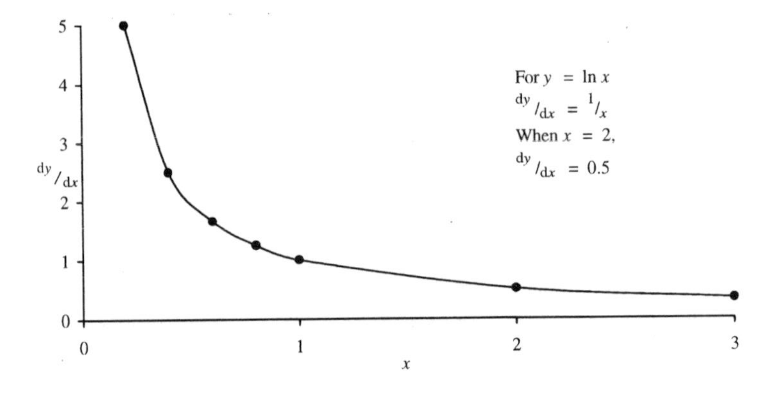

Graph 9.15

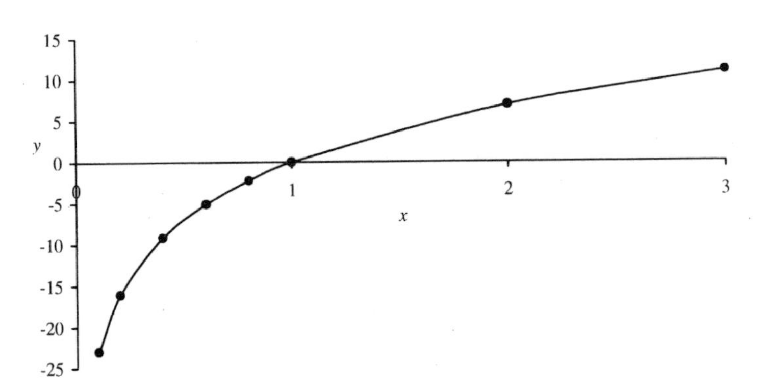

Graph 9.16

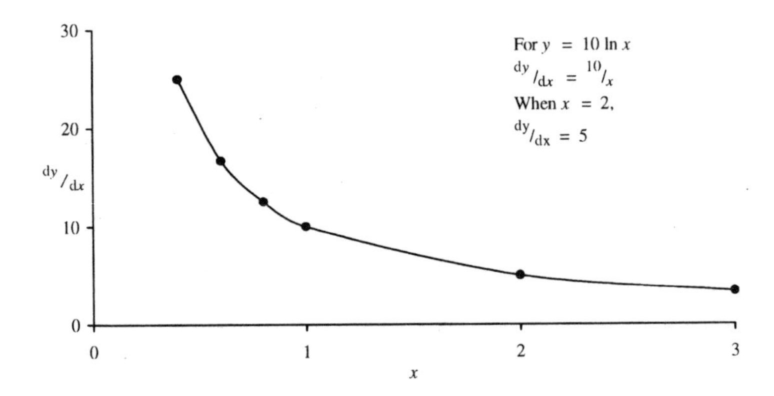

Graph 9.17

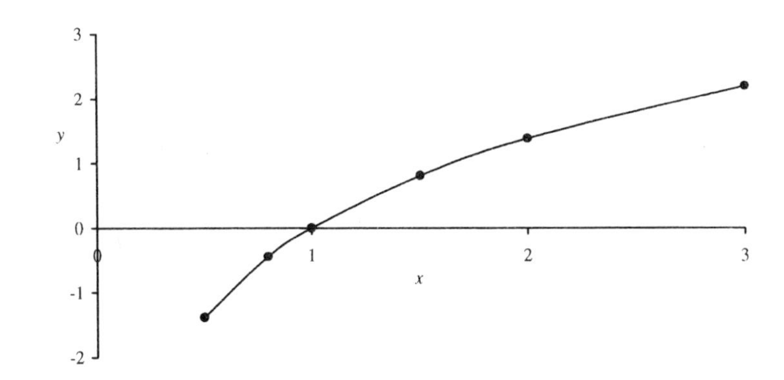

Graph 9.18

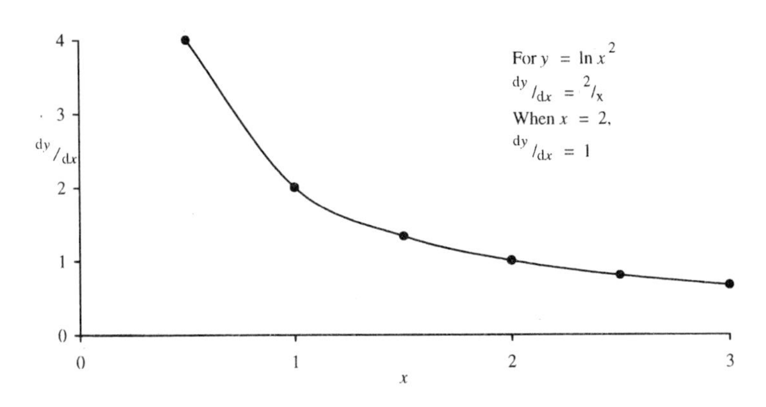

Graph 9.19

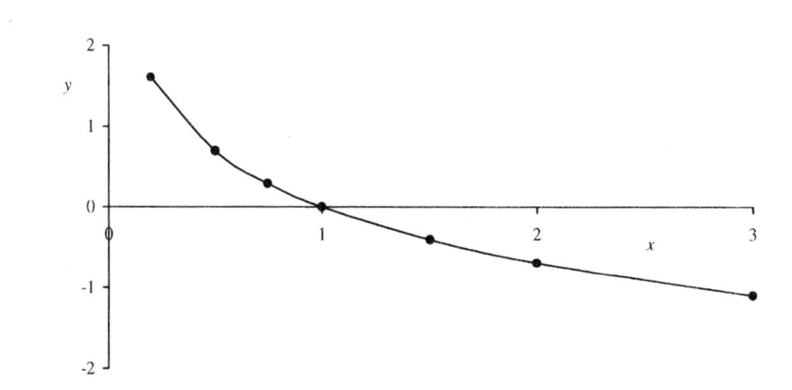

Graph 9.20

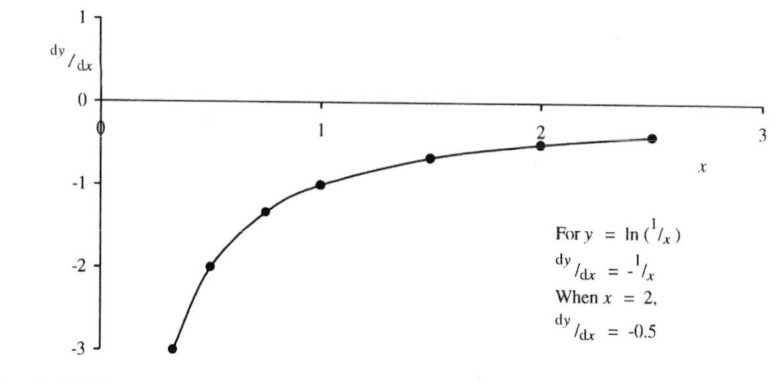

Graph 9.21

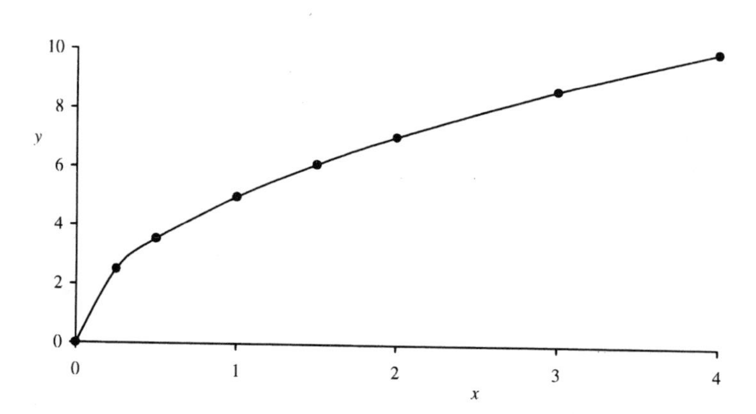

Graph 9.22

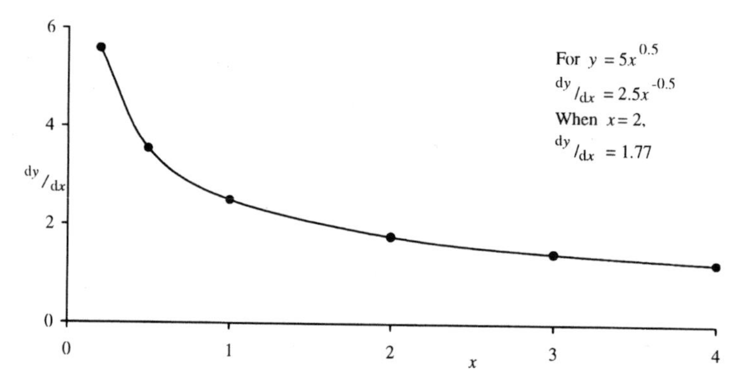

Graph 9.23

(ii) At $t = 2$ weeks = 14 days, ln dps = −0.433(14) + 19.3 = 13.24
therefore activity = antiln 13.24 = 5.6×10^5dps

13. Plot ln rate vs. $1/T$.
Gradient = $-2.16 \times 10^3 = -E_a/R$
therefore $E_a = 2.16 \times 10^3 \times 8.3$ J/mol = 17.9 kJ/mol
14. Plot $\log_{10}$ MR vs $\log_{10}$ BM.
Gradient = 0.77, y-intercept = 0.55
therefore log MR = 0.77 log BM + 0.55
therefore MR = $3.55 \times BM^{0.77}$
15. See Table 9.10
Rate $v = a[\text{substrate}]/(b + [\text{substrate}])$
Plot $^1/_v$ vs $^1/_{[S]}$; $^{[S]}/_v$ vs $[S]$; and v vs $^v/_{[S]}$
From these graphs , a = approx. 0.73 µmol/min, b = approx 60 µmol/1.

Table 9.10 Answers for Question 14, Chapter 8.

[Substrate]	*Rate*	*1/S*	*1/V*	*S/V*	*V/S*
0.063	0.37	15.9	2.70	0.17	5.87
0.084	0.42	11.9	2.38	0.20	5.0
0.105	0.46	9.52	2.17	0.23	4.38
0.14	0.51	7.14	1.96	0.27	3.64
0.25	0.58	4.0	1.72	0.43	2.32
0.35	0.63	2.86	1.59	0.56	1.80

INDEX